KU-308-213

General Preface to the Series

It is no longer possible for one textbook to cover the whole field of Biology and to remain sufficiently up to date. At the same time teachers and students at school, college or university need to keep abreast of recent trends and know where the most significant developments are taking place.

To meet the need for this progressive approach the Institute of Biology has for some years sponsored this series of booklets dealing with subjects specially selected by a panel of editors. The enthusiastic acceptance of the series by teachers and students at school, college and university shows the usefulness of the books in providing a clear and up-to-date coverage of topics, particularly in areas of research and changing views.

Among features of the series are the attention given to methods, the inclusion of a selected list of books for further reading and, wherever possible, suggestions for practical work.

Readers' comments will be welcomed by the author or the Education Officer of the Institute.

1977

The Institute of Biology,
41 Queens Gate,
London, SW7 5HU

Preface

Temperature is the single most important factor controlling plant distribution on a world-wide scale, and it also has a major effect on crop yields as has been illustrated recently by the disastrous consequences of the severe frost of June 1975 on coffee production in Brazil.

In view of its importance, it seems to me that insufficient emphasis is placed in biology courses in schools, colleges and universities on the effects of temperature on plants and this monograph is an attempt to arouse further interest in the subject. I know of no other place where the information collected here is summarized conveniently, and I hope that my study will be useful both to teachers and to students. There is still a lot to be learnt about the influence of temperature on plant growth and development and even simple experiments can yield unexpected and interesting results.

My thanks are due to the Biological Education Committee of the Institute of Biology, for the forbearance with which they have awaited my long-delayed manuscript; to numerous people who have supplied me with information; to Miss N. Browning and Mrs. B. Kilner who did the typing; to Messrs Fisons Ltd for supplying the photograph for Fig. 1–4; to those who have given me permission to use their published illustrations, and to my publishers, Edward Arnold Ltd for all their help.

Henfield, Sussex, 1977   J. F. S.

Contents

1 Temperature and Plant Life

1.1 Temperature

Temperature is an indication of the amount of heat energy in a system; the higher the heat content, the higher is the temperature. Like pressure, concentration and density, and in contrast to quantities such as volume and heat content, temperature is an *intensive* property of a system. So, if a homothermic material is divided into parts the temperature of each fragment is the same as that of the whole although the heat content is less. Temperature gradients determine the direction and rate of transmission of heat from one place to another; heat flows from a point of higher to one of lower temperature at a rate determined by the temperature difference and the thermal conductivity of the material.

Temperature is recorded on one of several scales relative to certain fixed points. The earliest scale was the one devised by the German physicist, Gabriel Fahrenheit in 1724. He divided the temperature range between a mixture of ice and water and supposedly-body temperature into a hundred point scale from 0–100 degrees. On this scale water freezes

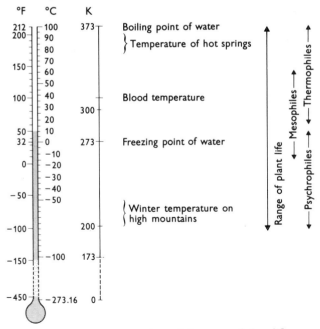

Fig. 1–1 Temperature scales and the range of plant life.

at 32 °F and boils at 212 °F (Fig. 1–1). The Fahrenheit scale is hardly ever used nowadays for scientific work, and for the sake of standardization it is also being replaced in everyday use, e.g. in meteorological reports, although it is in several ways more satisfactory for this purpose than the Centigrade scale.

In the Centigrade or Celsius scale, proposed by Anders Celsius, a Swedish physicist, in 1742, the temperature gradient between melting ice and boiling water at atmospheric pressure is divided into 100 degrees from 0–100. The units are therefore larger than those of Fahrenheit in the ratio of 180 : 100, i. e. 9 : 5.

Nowadays, in scientific work temperature is often measured on the 'absolute' scale devised by Lord Kelvin in 1848. The Kelvin scale starts at 'absolute' zero (0 K=−273.16 °C), the temperature at which the molecules in a system are at rest and the heat energy is therefore zero. The units are the same size as degrees Celsius and thus the absolute temperature can be calculated by adding 273.16 to the Celsius value (see Fig. 1–1).

The units and symbols used in this study are those of the International System of Units (SI) and a list of the SI units for various thermodynamic quantities is given in Table 1.

Table 1   SI units and symbols for various thermodynamic quantities, and some of their non-SI equivalents.

Quantity	Symbol	Unit	Symbol	Non-SI unit
temperature	T	Kelvin	K	$1\,°C = 1\,K$
				$1\,°F \simeq 0.55\,K$
heat	Q	joule	J	$1\,cal = 4.186\,J$
heat flux density	θ	watts per sq. m	$W\,m^{-2}$	$1\,cal\,cm^{-2}\,s^{-1}$ $= 4.186 \times 10^{-4}\,W\,m^{-2}$
thermal conductivity	λ	watts per m per Kelvin	$W\,m^{-1}\,K^{-1}$	$1\,cal\,cm\,s^{-1}\,K^{-1}$ $= 4.186 \times 10^{-2}\,W\,m^{-1}\,K^{-1}$
specific heat capacity	c	joules per kg per Kelvin	$J\,kg^{-1}\,K^{-1}$	$cal\,g^{-1}\,K^{-1}$
specific latent heat	l	joules per kg	$J\,kg^{-1}$	$cal\,g^{-1}$

1.2   Measurement of temperature

Temperature measurement depends on changes induced by temperature in a sensing device, e.g. changes in volume of a liquid, such as mercury in an ordinary thermometer, or differential expansion of a bi-metallic strip as in some kinds of thermograph. A precision of about 0.1–0.5 °C can be achieved with a sensitive thermometer or thermograph. Thermocouples are temperature-sensitive instruments in which electric current is caused to flow in a circuit consisting of two metallic junctions, usually of copper and constantan (Fig. 1–2). The amount of current flowing is proportional to the difference in temperature between the two junctions and if one is maintained at a constant temperature the

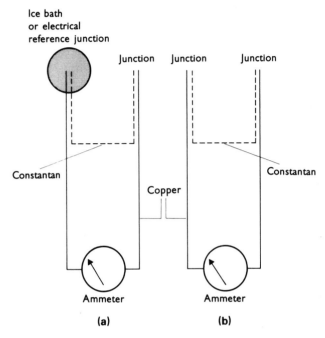

Fig. 1–2   Wiring for absolute (**a**) and differential (**b**) temperature measurement with thermocouples.

temperature of the other can be calculated from ammeter-readings (Fig. 1–2a). Alternatively, the difference in temperature between two junctions can be determined when both are varying (Fig. 1–2b). Thermocouples are small and relatively inexpensive to make but only produce a small electrical current which requires amplification before it can be recorded. They can be calibrated with an accuracy of 0.1–0.2 °C and are extremely versatile. They have been used extensively to measure the temperature of soil, water, air, plant tissues and surfaces and can also be adapted for measurement of radiant heat, e.g. in radiometers.

A thermistor is made of a semi-conducting crystalline material whose resistance to current flow varies with temperature. Thermistors are available with nearly linear temperature responses and they are extremely sensitive. Temperature changes of as little as 0.002 °C can be detected, but they are inherently less stable than thermocouples and require more frequent calibration.

1.3   The temperature of plants

Because of their relatively low heat production in relation to mass, and extensive surface area which is not well insulated against heat loss, plants are essentially poikilothermic or ectothermic organisms, i.e. their body temperature changes markedly with changes in external temperature. The inability of a plant to regulate its temperature is shown dramatically by loss of heat at night with the result that the temperature of the leaves may well fall well below air temperature; and by heating of the sunny side of a tree trunk to a temperature which may be as much as 30 °C higher than that on the shaded side on a cold day.

The heat generated by exothermic chemical reactions in plants is usually dissipated so quickly that it does not cause a significant rise in temperature, either of the tissues or their surroundings. However, when respiration is intense, and heat loss prevented, the temperature may rise appreciably. This can be demonstrated by germinating seeds in a vacuum flask and it is possible that even in the soil seeds may sometimes produce sufficient heat to raise their internal temperature, and thus hasten germination. The classic example of heat production in plants occurs in the fleshy inflorescence of plants of the arum family (*Araceae*). A rise in temperature of as much as 25 °C has been recorded inside the spathe of the wild arum, *Arum italicum*, when the inflorescence is mature and this is said to attract thermophilic pollinating flies to the flowers. Even inside the corolla of a flower such as a foxglove (*Digitalis purpurea*) temperatures significantly higher than in the surrounding air, have been detected. When the flower-buds of so-called 'snow' plants, e.g. the Alpine Soldanella, begin to grow in the spring, they may be covered by several centimetres of frozen snow. The growing buds are said to produce sufficient heat to melt the snow around them and form a channel through which the flower bud is carried on its stalk until it emerges and opens in the air above.

1.4   Effects of temperature on plant distribution

Because plants are not able to regulate their temperature effectively, their distribution, both on a world-wide and smaller ecological scale is strongly influenced by ambient temperatures. The world's climatic zones, determined largely by latitude and altitude, have characteristic vegetation types established by natural selection in response to the prevailing temperature and rainfall (Table 2). The influence of latitude on temperature is mainly due to the lowering in angle of the sun at mid-day with increasing distance from the equator, while the lower temperature at

Table 2   The World's climatic zones and natural vegetation. (Summarized from *The Times Concise Atlas of the World*. Times Books, 1975.)

Zone	Conditions	Typical natural vegetation
SUB–POLAR (Greenland, Iceland)	Winter very cold; summer short, cool and dry	tundra
COLD–CONTINENTAL (Canada, N. Russia)	Winter very cold; summer warm; rainfall small	boreal forest ('talgo')
SEMI–CONTINENTAL (N. and Central Europe)	Winter cold; summer hot; rain or snow throughout the year	mixed forest
WEST MARITIME (Western seaboards of Europe and U.S.)	Winter cool; summer warm; rain throughout the year	broad-leaved forest
EAST MARITIME (Eastern seaboard of U.S. and Canada)	Winter cold; summer warm; rain and snow heavier in winter	mixed forest
PRAIRIE STEPPE (Central U.S. and Russia)	Winter cool and dry; summer warm; rain mainly in spring	short grass (steppe)
MANCHURIAN (Manchuria)	Winter cold and dry, summer hot and generally wet	
HUMID–TEMPERATE (S.E. Asia, Brazil)	Winter warm and wet; summer hot and wet	sub-tropical forest
MEDITERRANEAN	Winter mild with moderate rain; summer warm and dry	scrub
SEMI–ARID (Middle East, parts of Africa, U.S., Australia)	Winter cool or warm, usually dry; summer hot with variable rainfall	deserts in dryer places
DESERT (N. Africa, part of Australia)	Considerable range of temperature through the year; no regular rainfall	desert
DRY TROPICAL (India, Central South America	Winter hot and dry; summer hot and wet	savannah, monsoon forest
WET TROPICAL	Hot with heavy rainfall throughout the year	tropical rain forest (selva)
MOUNTAIN (Andes, Himalayas, etc.)	Low temperatures predominate	mountain

high altitude is attributable largely to the expansion of warm air as it rises from below causing it to cool. Roughly speaking, an increase in altitude of 160 m entails a fall in the mean temperature of about 1 °C.

Fig. 1–3 shows the mean monthly temperature throughout the year on the summit of Ben Nevis (1343 m above sea level), the highest mountain in Scotland, and at nearby Fort William (0.9 m above sea level). The variation in temperature between winter and summer is similar in the two

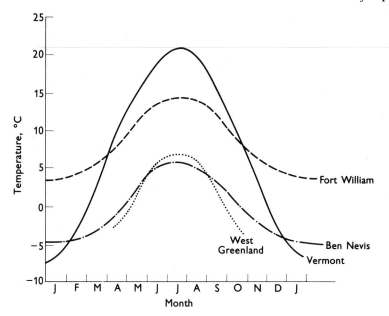

Fig. 1–3 Mean monthly temperatures at Fort William and at the summit of Ben Nevis in Scotland; in West Greenland; and in Vermont (New England, U.S.A.). (Redrawn from PEARSALL, 1950.)

places but the mean temperature for each month is lower on Ben Nevis by about 8 °C throughout the year. Even in mid-summer (June–July) the mean monthly temperature at the summit of Ben Nevis hardly reaches the winter temperature at Fort William and is similar to that of West Greenland (latitude 65° N) in summer. Obviously plants can only survive at high altitude if they are adapted to grow at low temperature or have a very short growing season.

In spite of their modest height, the mountains in Great Britain become treeless at a relatively low level and trees are rarely found above 600 m. In contrast in New England, U.S.A., where the mean annual temperature in the lowlands is about the same as in Scotland, the tree line, e.g. on Mount Washington (1917 m) occurs at about 1500 m. The explanation of this lies in the fact that the summer temperature in New England is higher than in Scotland both in the lowlands and at the same altitude on mountains. Thus at a height corresponding to the summit of Ben Nevis the mean July temperature on Mount Washington is 12 °C and the growing season is much longer. Growth of trees at high altitudes on British mountains is evidently limited by low summer temperatures rather than by winter temperatures which are higher than those on Mount Washington at the same height. It should be noted that although air temperatures are quoted here it is likely that soil temperatures are more important in limiting the growth of plants on mountains and these remain significantly lower than air temperature throughout the summer in Scotland restricting absorption of water and nutrients and root growth.

In other situations plant distribution may be controlled by winter temperature and in particular by the severity of frosts. Frost-sensitive (Chapter 5) and chilling-sensitive plants (Chapter 3) are limited in their distribution by the minimum temperature to which they are exposed, even briefly, during the year. On the other hand, many plants are unable to survive in deserts because of their inability to withstand the maximum temperatures occurring during the hottest periods. The geographical distribution of the stemless thistle, *Cirsium acaule*, in Europe appears to be largely determined by temperature. From Sweden to south-western England the distribution follows closely the isotherms for mean and maximum temperatures in July, August and September, suggesting that summer temperature is important in determining its western limits.

Every plant has an optimal temperature for growth and upper and lower limits beyond which all growth stops (Chapter 4). Not only do species differ from one another but even different strains of the same species may have different temperature tolerances. Ecotypes of widely-distributed species sometimes, but not always, have different temperature optima, and these have presumably been established by natural selection to suit the prevailing conditions.

Establishment of an optimal temperature for growth is often complicated by the fact that plants do not normally grow under constant temperature conditions even in the tropics, and many grow better at a fluctuating than at a constant temperature. This is an example of thermoperiodism which will be discussed further in Chapter 4.

The temperature range over which plants will survive is usually considerably wider than that within which they can grow and complete their life cycle. An alpine willow, *Salix pauciflora*, may withstand freezing to $-70\ ^\circ\text{C}$ in winter, but will not grow or even survive unless the temperature rises much higher than this in summer (p. 41). The distribution of the moor-rush, *Juncus squarrosus*, on British mountains is limited by its failure to produce viable seeds above about 900 m although the plants will grow vegetatively up to 1200 m.

1.5   Controlled environments

The main purpose of growing plants in 'hot'-houses, conservatories or glasshouses is to maintain an optimum temperature for growth and protect susceptible crops from frost damage. In this way the number of places in which such crops can be grown is increased and the season can be extended. The ever-increasing cost of fuel for heating makes it necessary periodically to re-appraise the optimal temperature for commercial glasshouses as this must be determined by economic return on investment rather than by maximum yield.

Growing plants under glass presents a number of problems. One is that light intensity is reduced; even modern glasshouses when clean rarely have a light intensity inside higher than 70% of that outside. Another difficulty is that heat generated from solar radiation (see Chapter 2) is trapped in a glasshouse because of the opacity of glass to infra-red radiation (the so-called 'glasshouse effect'), and so the plants may get over-heated during the day unless adequate ventilation is provided.

The high capital cost of glasshouses has led to experimentation with cheaper materials, e.g. thin plastic film for the protection of crops. This suffers from a number of disadvantages including low light transmission, short life, high maintenance costs and problems of ventilation; the use of such materials is still largely confined to areas where protection is only marginally necessary, e.g. in parts of Italy, Japan and Israel. In the U.K. plastic film is mainly used to cover strawberries and salad crops such as lettuce and celery to protect them from late frosts.

Because temperature has such a profound effect on plants it is usually necessary for plant physiologists to control the temperature during their experiments. If the temperature is kept constant the effects of other environmental factors can be studied, and by varying the temperature in a controlled manner the interaction between temperature and such factors as light intensity can be investigated. Considerable insight into the mechanism of photosynthesis (see Chapter 3) and the control of plant growth (Chapter 4) has been obtained in this way.

The degree of control of temperature (and light intensity) required by a plant physiologist is often much greater than that which can be achieved in a glasshouse, and sophisticated 'phytotrons' have been designed for this reason. They range from extensive suites of rooms maintained at a variety of different temperatures, to portable chambers the size of refrigerators, which can be set to a pre-determined temperature (Fig. 1–4). In the best of such equipment temperature can be controlled to within 1 °C irrespective of the light intensity. Time-clocks are incorporated to enable temperature and/or light intensity to be varied periodically and independently of one another. The major problem in designing phytotrons is to maintain sufficient air-flow to reduce variation in temperature between different parts of the chamber to an acceptable minimum.

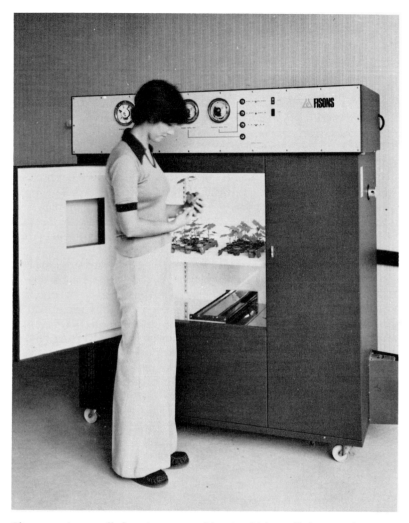

Fig. 1–4 A controlled-environment cabinet in which small plants can be grown at a constant temperature. (Photograph reproduced by kind permission of Fisons Scientific Apparatus Ltd, Loughborough, Leicestershire.)

2 Heat Balance in Plants

2.1 The plant as a heat exchanger

According to the first law of thermodynamics, energy cannot be created or destroyed, but only changed from one form to another. Heat energy is acquired by a plant either directly, e.g. by convection and conduction, if the plant is cooler than the surrounding air; or indirectly by transformation of chemical energy in exothermic metabolic processes such as respiration, and by conversion of radiant energy, to heat. On the other hand, heat is lost by convection and conduction when the plant is at a higher temperature than its surroundings, by the latent heat of evaporation of water, by uptake of water at a lower temperature than itself and by emission of long-wave (infra-red) radiation.

When the loss of heat exceeds acquisition the temperature of the plant falls, and vice-versa, and depending on the conditions, a balance is established at a temperature which may be somewhat above or below the ambient. Different parts of plants, e.g. roots and leaves, are usually at different temperatures because of the different conditions to which they are exposed. Temperature gradients thus exist in the plant but flow of heat along such gradients is relatively slow because of the low thermal conductivity of water and plant tissues. As most of the energy that is exchanged between a plant and its environment occurs through the leaves attention is focused on them in the discussion which follows.

The energy balance of a leaf (or any other plant part) can be represented by the equation:

$$Q_a = Q_r + Q_e + Q_c + Q_m + Q_s$$

where

Q_a = total incident radiation absorbed;

Q_r = energy re-radiated as infra-red;

Q_e = latent heat of evaporation;

Q_c = energy exchanged between leaf and air by conduction and convection; conduction is the transfer of heat between molecules through collisions (cf. diffusion), while convection involves movements of groups of molecules of a gas or liquid as a result of pressure gradients (see p. 14). This term also includes any exchange of heat between the leaf and other parts of the plant, e.g. via the transpiration stream;

Q_m = energy balance between endothermic reactions such as photosynthesis and exothermic reactions such as respiration. When the latter exceeds the former Q_m is positive and when it is lower, Q_m is negative. While Q_m is of great physiological significance because this is the part of the energy budget upon which growth depends, the amount of energy involved is relatively small (less than 1% of Q_a);

Q_s=energy stored; when more energy is entering the leaf than is leaving it, i.e.

$$Q_a > Q_r + Q_e + Q_c + Q_m$$

Q_s is positive and the temperature of the leaf rises and vice versa. When losses balance gains, Q_s is zero and the temperature remains steady.

2.2   Absorption of solar radiation

Sunlight may fall on a leaf either directly, after scattering by molecules in the air ('sky-light'), or after reflection from clouds ('cloud-light') or other surrounding objects. The amount of solar radiation incident on the earth's atmosphere, the so-called 'solar-constant', is about 8.4 J per cm² per min and of this about 50% reaches the earth's surface at sea level on a cloudless day when the sun is directly overhead. Light intensity reaches a maximum of about 10 000 lux under these conditions.

In addition to solar radiation, infra-red or thermal radiation is also absorbed by a leaf. Any object at a temperature above 0 K emits long wavelength radiation, mainly in the infra-red (> 1500 nm). Most of the thermal radiation from the sky comes from water, CO_2 and other molecules in the atmosphere but surrounding objects also contribute a quota. The amount of radiation absorbed by a leaf depends on the wavelength and on its spectral characteristics. Fig. 2–1 shows the amounts of radiation of different wavelengths absorbed, transmitted and reflected by a green leaf. Visible light is absorbed most effectively via the chlorophyll pigments in the blue (400–500 nm) and red (600–700 nm) regions of the spectrum. Green light (500–600 nm) is absorbed to a lesser extent and there is a corresponding increase in the amount of light transmitted and reflected in this region with a peak at about 550 nm. Leaves absorb weakly in the near infra-red (700–1500 nm) and since this accounts for a substantial proportion of the total energy in solar radiation the quantity of potential heat energy absorbed is much reduced. On the other hand, plants are efficient absorbers of far infra-red (1500–10 000 nm), but since solar radiation is weak in these wavelengths after its passage through the atmosphere, heat gain from this region of the spectrum is relatively small.

Thermal radiation is also emitted by plants and so vegetation can be photographed in the dark on infra-red sensitive film. A large fraction of the energy derived from absorption of visible radiation is actually re-radiated as infra-red.

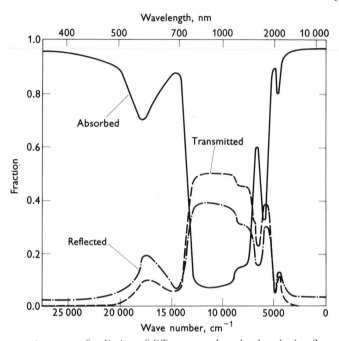

Fig. 2–1   Amounts of radiation of different wavelengths absorbed, reflected and transmitted by a leaf.

Some plants, notably desert plants, have silvery or shiny leaves which increase the amount of solar radiation reflected. Reflectance at 50 nm is increased in some such leaves from the usual 10–20% to 30% of incident radiation with a corresponding reduction in energy absorbed. Such a reduction has a marked effect on leaf temperature and may be important in situations where leaf temperature tends to rise above the optimum for photosynthesis.

It is well known that tender plants are more prone to frost injury on clear than on cloudy nights. This is because the clear sky acts as a 'black body' which absorbs heat energy from the leaves in the form of infra-red radiation causing their temperature to fall. Clouds reflect this radiation back and cooling is reduced. The alteration in position, or folding of leaves, such as those of *Acacia*, at night may be devices for reducing radiation cooling rather than transpiration (see Chapter 4).

2.3   The influence of transpiration

The evaporation of water requires energy, and the latent heat of evaporation is so high (2450 J g^{-1} at 20 °C) that transpiration may dissipate as much as 50% of the total energy absorbed by a mesophytic leaf under conditions of high transpiration. GATES and his co-workers at the Missouri Botanical Garden have made a detailed study of the relationship between transpiration rates and leaf temperature over a wide range of experimental conditions. Fig. 2–2 shows the results of some of their

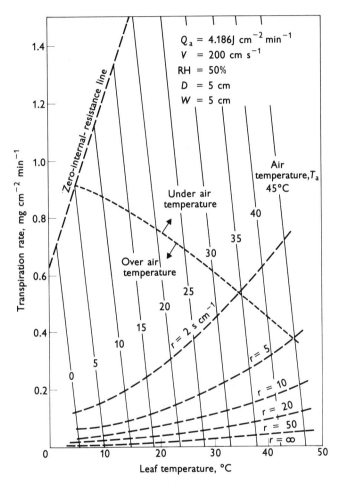

Fig. 2–2 Relationship between transpiration rate and leaf temperature at various air temperatures (T_a) and diffusive resistances (r). (From GATES, 1968.)

calculations based on computer simulations of the behaviour of a hypothetical leaf. The leaf in question is assumed to have a surface area of 25 cm² (D × W) and to be absorbing energy at the rate of 4.186 J cm⁻² min⁻¹ (Q_a). Atmospheric humidity is taken to be 50% (relative humidity,

RH) and air moving over the leaf at 200 cm s^{-1} (V). These conditions correspond approximately to average conditions in temperate latitudes on a clear sunny day when there is a light breeze.

The relationship between transpiration, leaf and air temperature was examined at various sizes of stomatal aperture (diffusive resistance, r from 2 cm^{-1} to ∞). With increasing leaf and air temperature transpiration rate increases, particularly when the stomata are wide open. Under the set of conditions considered here, leaf temperature is always higher than air temperatures when the temperature of the latter is below 35 °C, but above this value the leaf is cooler than the air if transpiration is sufficiently high.

Other calculations have shown that the temperature of a large leaf rises much more under conditions of high light intensity than does that of a small leaf and this may account for the prevalence of small-leaved species among desert plants. Direct measurements on small-leaved desert species, e.g. the sage-brush, *Artemisia tridentata*, confirms that leaf temperature remains close to ambient over a wide range of air temperatures and such plants are clearly at an advantage in being able to avoid reaching lethal temperatures without excessive transpiration. In contrast, in cacti, e.g. *Opuntia* spp. where the leaves are reduced to scales and the stem is the photosynthetic and transpiring organ, surface temperature rises markedly above the ambient on occasions and such plants obviously tolerate high temperatures rather than avoid them.

It is often asserted that a 'function' of transpiration is to prevent leaves from becoming too hot. From what has been said above it should be clear that transpiration does reduce leaf temperature, but its effect is most marked in mesophytes with low stomatal resistance, i.e. in plants that transpire rapidly. In xerophytes, stomatal resistance is generally high and transpiration rates correspondingly low. Such plants cannot afford to use large quantities of water for cooling and either must depend on other methods of heat dissipation such as re-radiation or acquire high temperature tolerance (see Chapter 6).

2.4   Conduction and convection

Heat is conducted between a leaf surface and the surrounding air by the random thermal collisions of gas molecules with the leaf surface and the amount of heat transferred depends on the temperature gradient and thermal conductivity at the leaf/air interface. Convection on the other hand involves turbulent movement of a fluid or gas brought about by changes in density or pressure induced by temperature gradients. 'Free' or 'natural' convection occurs when heat dissipated by a leaf causes air round the leaf to expand and so decrease in density. This air rises and is replaced by cooler air from below. Wind can also remove heated air from the leaf surface and this is called 'forced' convection. Even at very low wind speeds (say 10 cm s^{-1}) forced convection is much more important than free convection in causing loss of heat from a leaf.

It has been found in experiments in wind-tunnels with thin copper plates of various shapes that regardless of wind speed or orientation a disc-shaped plate has the lowest heat dissipation per unit of surface area while an elongated or lobed structure has a significantly higher rate of convective cooling under all conditions. The common occurrence of needle-shaped and deeply-lobed leaves, especially in desert plants, and the rarity of circular leaves is therefore understandable in terms of maximum heat dissipation.

2.5　Heat balance in plant communities

The temperature relations of a plant community are immensely more complex than those of a single leaf because of changes in orientation of leaves during growth, mutual shading, and complicated patterns of air flow and humidity gradients within the leaf canopy, so only broad generalizations can yet be made.

LEMON (1963) has estimated that for a plant community, such as grassland, consisting of short dense vegetation, well supplied with water

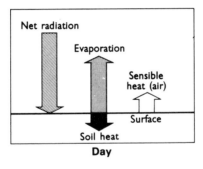

Day

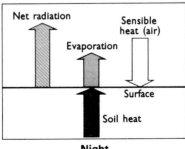

Night

Fig. 2–3　Diagrammatic representation of the energy budget at the earth's surface during the day and at night. (From TANNER and LEMON, 1962.)

and nutrients, 75–85% of the incoming radiation during the day is dissipated in the evaporation of water from leaf or soil surfaces, 5–10% is absorbed by the soil, a similar amount transferred to the air by conduction and convection, and not more than about 5% utilized in photosynthesis. Evaporation is also responsible for a significant heat loss at night but re-radiation is even more important (Fig. 2–3). The presence of vegetation reduces the absorption of heat by soil during the day and suppresses heat loss at night.

Vegetation at high altitudes, e.g. alpine tundra, presents a somewhat different picture to that at lower altitudes. Two factors that are of particular importance are the increase in light intensity and decrease in temperature. Radiation heating is thereby increased during the day as also is the rate of heat dissipation. Plants at high altitudes are particularly susceptible to injury by excessive cooling at night. Plants of the arctic tundra are exposed to somewhat similar conditions as those at high altitudes and this no doubt accounts for some of the resemblances between the two kinds of vegetation, which in general consist of low-growing, woody shrubs with relatively small leaf area.

3 Temperature and Metabolism

3.1 Temperature coefficient (Q_{10})

A common way of assessing the effect of temperature on a biological process is to calculate the value of Q_{10}, or temperature coefficient. Q_{10} is the ratio of the rate of a process at one temperature to that at a temperature 10 °C lower, i.e.

$$Q_{10} = \frac{K_2}{K_1}$$

where
K_2 is the rate at T °C
K_1 is the rate at $T-10$ °C

It is not always convenient to measure the rates at intervals of exactly 10 °C, and approximate values for Q_{10} can be calculated from measurements at any two temperatures using the equation:

$$Q_{10} = \left(\frac{K_2}{K_1} \right)^{\frac{10}{T_2 - T_1}}$$

where
K_2 is the rate at the higher temperature, T_2
K_1 is the rate at the lower temperature, T_1

Over the range of temperature at which most living organisms exist (5–40 °C) the rate of a chemical reaction is roughly doubled for every rise of 10 °C, i.e. $Q_{10} \simeq 2$. On the other hand, physical processes, e.g. diffusion, and photochemical reactions have a low Q_{10} ($\simeq 1.2$). By examining the effect of temperature on photosynthesis at different light intensities and carbon-dioxide concentrations, F. F. BLACKMAN (1905) deduced that photosynthesis involves a light phase with a low Q_{10}, and a dark phase with a high Q_{10} in which CO_2 is reduced. Alteration of temperature over a limited range has little effect on photosynthesis at low light intensities when the light phase is limiting and a large effect when light intensity is high (see Fig. 3–1).

In living organisms over the physiological range of temperature, enzyme-catalysed reactions have a Q_{10} similar to that of chemical reactions. However, at temperatures above about 40 °C many enzymes become inactivated by denaturation of their proteins and there is a progressive reduction in the rate of the reaction with time. Thus, while the

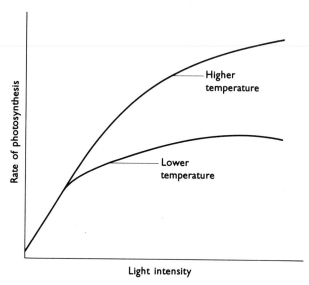

Fig. 3–1 Relationship between light intensity and photosynthesis at two different temperatures.

Q_{10} may be high initially it falls rapidly and may eventually reach values lower than unity. It must be pointed out however that some enzyme proteins are thermostable, and a convenient way of separating α and β amylases, for example, is to heat the enzyme mixture to about 80 °C when the α amylases are inactivated but the β amylases are unaffected.

Physiological processes such as respiration which are controlled by enzymes are affected by temperature in a similar way. In pea seedlings, for example, the rate of oxygen absorption increases with temperature up to about 35 °C and Q_{10} values between 2.0 and 2.5 are commonly observed. Above this temperature the rate of respiration is increased initially but falls markedly with time, and the effect is more marked the higher the temperature (Fig. 3–2).

3.2 The Arrhenius equation

The effect of higher temperatures on a chemical reaction cannot be explained solely in terms of an increase in kinetic energy causing more collisions between molecules. It has been calculated that a 10 °C rise in temperature causes only about a 2% increase in the number of collisions and this would not be expected to double the rate of the reaction. In 1889, a Swedish chemist, Svante Arrhenius, suggested that raising the temperature by 10 °C doubles the number of molecules which have a sufficiently high energy content (activation energy) to undergo reaction.

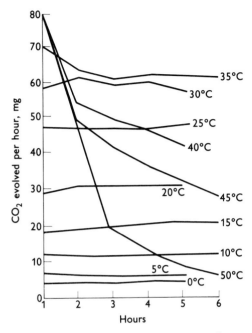

Fig. 3–2 Effect of temperature on the respiration rate of pea seedlings. (After KUIJPER. Redrawn from HAAS and HILL, 1929.)

He showed that the effect of temperature on the rate of hydrolysis of sucrose could be represented by the equation:

$$\frac{d \ln K}{dT} = \frac{E}{RT^2}$$

where

K is the rate of the reaction
R is the gas constant (8.314 J mol⁻¹ K⁻¹)
T is the absolute temperature (K)
E is the activation energy (J)

Integration of this equation gives:

$$\frac{\ln K_2}{K_1} = \frac{E}{R} \left(\frac{1}{T_1} - \frac{1}{T_2} \right)$$

where K_2 and K_1 are the rate constants corresponding to T_2 and T_1 respectively.

From this it can be seen that the value of E can be obtained by measuring the slope of the straight line obtained when log K is plotted against $1/T$ (Fig. 3–3). Thus by measuring the rate of a reaction at several tempera-

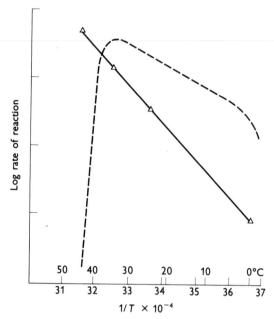

Fig. 3-3 Arrhenius plot for the decomposition of nitrogen pentoxide (Δ—Δ) and for an enzyme reaction (– – – – –).

tures, E can be obtained, and knowing E and the rate of one temperature, the rate at another temperature can be calculated. Values of E are such that over a limited range of temperature:

$$\frac{K_2}{K_1} \simeq 2 \quad \text{i.e.} \quad Q_{10} \simeq 2$$

where

K_1 is the rate of the reaction at a temperature T_1 °C
K_2 is the rate of the reaction at a temperature of $T_1 + 10$ °C.

Most enzyme reactions follow the Arrhenius equation over a certain range but at temperatures close to zero and above about 40 °C the relationship breaks down due to inactivation of the enzyme (Fig. 3-3). Arrhenius used the term 'temperature characteristic' in place of activation energy for complicated biological processes such as respiration. The Q_{10} of respiration falls more rapidly with increasing temperature than would be expected on the basis of the Arrhenius equation. One possible explanation of this is that whereas at low temperature the rate is controlled by a partial process, e.g. an enzyme reaction, having a high Q_{10}, at higher temperatures it may be limited by one having a low Q_{10}, e.g. the rate of diffusion of oxygen into the tissue (cf. photosynthesis, p. 17).

3.3   Chilling injury

Whereas most enzyme-catalysed reactions give a linear Arrhenius plot over the physiological temperature range there are a number of cases in which a plot of the logarithm of velocity against the reciprocal of absolute temperature shows two straight lines of different slope, i.e. there is an abrupt change of E at a particular temperature. This phenomenon has been observed most often with enzyme systems associated with membranes, e.g. in mitochondria, and occurs in those plants, usually of tropical or sub-tropical origin, that are injured by low non-freezing temperatures (Fig. 3–4). Chilling injury commonly takes place at

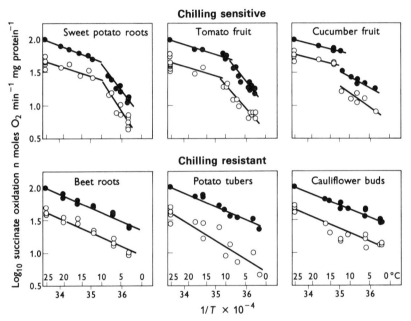

Fig. 3–4   Arrhenius plots of succinate oxidation by isolated plant mitochondria. (From LYONS and RAISON, 1970.)

temperatures of 10–12 °C and this is about the point at which the discontinuity occurs in the Arrhenius plot for selected enzymes. Physiological processes such as respiration show a similar discontinuity in Arrhenius plots at the critical temperature in chilling-sensitive species (Fig. 3–5).

Chilling disrupts the entire physiology of sensitive plants and a number of mechanisms have been proposed to account for the effects. It seems likely that the cell membranes in chilling-sensitive plants undergo a physical phase transition at the critical temperature from a normal

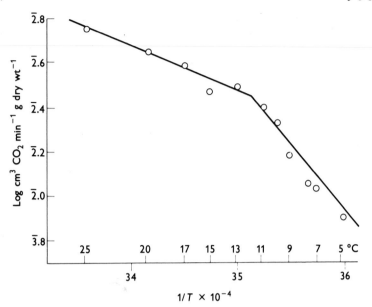

Fig. 3–5 Arrhenius plot of the respiration rate of cucumber leaves. (Redrawn from MINCHIN and SIMON, 1973.)

flexible liquid crystal to a solid gel structure (Fig. 3–6). This change in state would be expected to bring about a contraction of the membrane components causing the formation of holes and increased permeability.

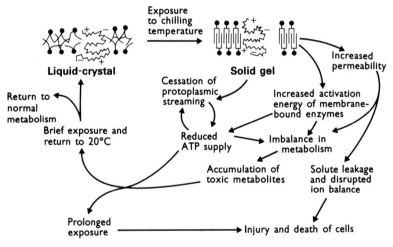

Fig. 3–6 Diagrammatic representation of the events leading to chilling injury in sensitive plant tissues. (From LYONS, 1973.)

The phase transition may also increase the activation energy of membrane-bound enzymes leading to interference with metabolic processes. Reduction in the supply of ATP coupled with increased rigidity of membranes may cause cessation of protoplasmic streaming, which is one of the immediate symptoms of chilling injury.

Unlike freezing injury (see Chapter 5) exposure to chilling temperatures often needs to be prolonged (sometimes several weeks is necessary) before injury occurs. However, there are exceptions; the leaves and ripening fruits of banana for example are damaged by brief exposure to temperatures below 10 °C. In general, the degree of chilling injury in sensitive plants increases as temperature is lowered or exposure time extended at any temperature below the critical one. In some cases it has been observed that the effects of exposure to chilling temperature are ameliorated if the plant is subsequently transferred to a higher temperature. In *Zea mays* seedlings visible leaf injury occurs within 36 hours of exposure to 3 °C but upon transfer to 21 °C the leaves return to normal and injury symptoms disappear within 72 hours. However, chilling-sensitive plants cannot apparently be 'hardened' by gradual exposure to lower temperatures as in the case of frost-sensitive plants (see Chapter 5).

Analyses have shown that there is a higher proportion of saturated fatty acids in the lipids of chilling-sensitive than of chilling-resistant plants. A similar difference in the proportion of saturated to unsaturated fatty acids has also been found in the lipids of homeothermic as compared with poikilothermic animals and it is believed that membranes containing a high proportion of saturated fatty acids are less stable and more liable to undergo the phase transition at chilling temperatures.

3.4   The biological clock

All living organisms seem to possess an in-built time-keeping mechanism by means of which their daily and seasonal activities are regulated. Some examples of the operation of the biological clock in relation to plant growth and thermo-periodic phenomena will be mentioned below (Chapter 4). A time-keeping mechanism which ran faster (or slower) at higher temperatures would be quite ineffective and so, like man-made clocks and watches, biological clocks are temperature-compensated, i.e. they have a low Q_{10}. Values of Q_{10} for some of the diurnal (circadian, i.e. about 24 hours) rhythms occurring in plants are given in Table 3.

Table 3 shows that while most rhythms run faster, i.e. have somewhat shorter periods at higher temperatures ($Q_{10} > 1.0$), some actually run slower. When the periodicity of a rhythm is plotted against temperature there is a peak (or trough) at a certain temperature (see Fig. 3–7). This kind of curve suggests that temperature insensitivity is the result of an

Table 3   Temperature dependence of various plant rhythms (from SWEENEY, 1969).

Plant	Rhythm	Q_{10}
Neurospora sp.	Zonation in dim red light	1.03
Pilobolus sp.	Sporulation	1.30
Gonyaulax polyedra	Luminescent glow in the dark	0.90
(a dinoflagellate)	Cell division in dim light	0.85
Oedogonium sp.	Sporulation in light	0.80
Euglena sp.	Phototaxis	1.01–1.10
Helianthus annuus	Root exudation	1.10
Avena sativa	Coleoptile growth	1.00
Phaseolus vulgaris	Sleep movement of leaves	1.01–1.30

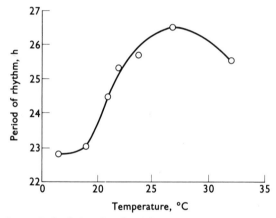

Fig. 3–7   The period of the circadian rhythm in luminescence in *Gonyaulax polyedra* as a function of temperature. (From SWEENEY, 1969.)

interaction between two or more reactions having different temperature coefficients. Let us suppose that the effect of one reaction is to increase the period and of another to decrease it. If both reactions are affected in exactly the same way by a change of temperature the period will not change and perfect temperature compensation will be achieved. If on the other hand the two reactions are differently affected, the period might either lengthen or shorten with increasing temperature depending upon which reaction was affected most. The fact that Q_{10} values lower than unity can be observed seems to rule out the possibility that the biological clock is controlled entirely by a temperature–insensitive physical process, such as diffusion, but it is possible that physical as well as chemical reactions may be involved in a temperature–compensated system such as that described above.

4 Temperature and Growth

4.1 Reponse to temperature

Because temperature affects metabolism, it also influences growth. Whether measured as dry weight increase or in any other way, growth is slow at temperatures close to 0 °C and increases rapidly with temperature up to maximum value which for mesophilic plants is commonly in the range of 20–35 °C (Fig. 4–1). Beyond this the rate of growth begins to level out and then declines until a temperature is reached at which growth stops. At a still higher temperature than this death occurs.

The optimal temperature for growth and the range over which it occurs varies widely between species and even between ecotypes of the same species. In general, plants that are natives of colder regions have lower temperature optima and lower death temperatures than those of warmer places, but can survive better at lower temperatures. Some green algae found in arctic and antarctic waters survive indefinitely, and presumably grow, at temperatures which rarely exceed 0 °C by more than a fraction of a degree. At the other extreme blue-green algae found in hot-springs can grow at temperatures at least as high as 85 °C.

The typically asymmetric bell-shaped growth curve shown in Fig. 4–1 is

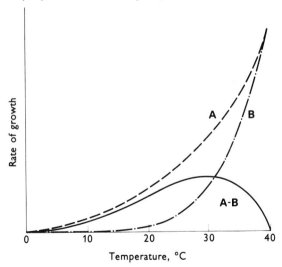

Fig. 4–1   Effect of temperature on the rate of growth of a mesophilic plant. The growth curve (**A–B**) is the resultant of curve **A** which represents the increase in activation energy of chemical reactions with increasing temperature and curve **B** which represents the increasing inactivation of enzymes as temperature rises.

thought to be the resultant of two opposing influences. On the one hand the increased activation energy of reacting molecules with increasing temperature (curve **A**) favours increased growth, and on the other hand, the increased inactivation of enzymes with rising temperature (curve **B**) depresses it. The relationship is similar to that between temperature and other physiological processes but the correlation is not exact. Respiration, for example, usually has a higher optimal temperature, while that of photosynthesis is commonly lower, than that of growth.

Growth responses to temperature are actually much more complicated than this rather simple analysis indicates. The optimum temperature for growth varies between different organs on the same plant and even between two sides of the same organ (see p. 38 below). It also changes as a plant ages so that the apparent optimum may vary according to the length of time over which the measurements are made (cf. temperature effects on respiration, p. 18). Changes in the temperature of one part of a plant, say the roots, may affect growth of another (e.g. the shoot) even when the temperature of the latter is kept constant (Fig. 4–2).

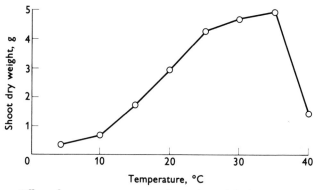

Fig. 4–2   Effect of root temperature on shoot dry weight in *Zea mays*. (Redrawn from COOPER, 1975.)

The effects of temperature on growth of plants under natural conditions is complicated by the large diurnal and seasonal fluctuations that occur. Estimates have been made of the number of heat units (temperature × time) required to reach various stages in the development of a plant and tables of so-called 'thermal constants' have sometimes been published. These are usually only a rough guide to the temperature requirements of a species as the pattern of temperature variation is often of great importance, and light intensity contributes greatly to the heat content of a plant (see Chapter 2). However when light fluxes are fairly constant the accumulative effects of temperature during the season may determine the time taken for a crop to reach maturity. This appears to be the case, for example, with pineapples growing in Hawaii.

Everyone who has any experience of growing plants in a greenhouse or the sitting-room knows that many grow better when the night temperature is lower than day temperature. Tomatoes, for example, grow faster when the temperature is 26 °C by day and 17 °C at night than they do at a constant temperature of 26 °C or any intermediate temperature (Fig. 4–3). For this reason tomatoes do not grow well in warm countries

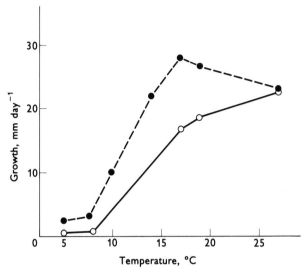

Fig. 4–3   Growth in height of tomato plants at a range of constant temperatures (o———o), and with a constant day temperature (26 °C) and range of night temperatures (●–––●). (Redrawn from WENT, 1945.)

unless the temperature falls appreciably at night. Professor F. W. WENT of the University of California has demonstrated that the geographical distribution of some plants may be determined by diurnal fluctuation of temperature during the growing season. The changes in the mean monthly day and night temperatures at Pasadena, California, throughout the year are represented by the narrow ellipse in Fig. 4–4. *Zinnias* have a range of temperature requirements indicated by the dotted circle labelled 'Zinnia', and since such conditions exist in Pasadena from July to September, the plants grow well there during these months. China Asters require lower day and night temperatures than *Zinnia* and these conditions occur in spring and autumn. In contrast *Saintpaulia* needs a higher night temperature than occurs in Pasadena and so it is not found there. A small annual composite, *Baeria charysostoma*, found in mountain valleys and foothills in California cannot tolerate high night temperatures and is killed if the temperature exceeds 26 °C although it flourishes at day temperatures considerably in excess of this. Why plants such as this can

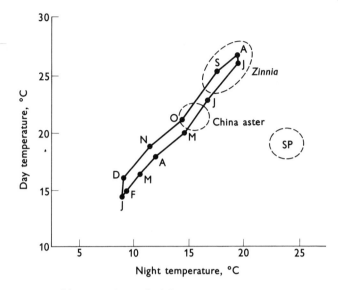

Fig. 4–4 Monthly mean day and night temperatures in Pasadena, California, and the temperature ranges of *Zinnia*, China Aster and *Saintpaulia* (SP). (Redrawn from WENT, 1957.)

tolerate higher temperatures by day than at night is unknown (see Chapter 6).

It has been found that low night temperature causes some plants to have a higher sugar content and this may be one of the factors promoting increased growth under these conditions. Reduction in night temperature leads to increased root growth relative to shoot growth in tobacco and potatoes and transport of sugars to the roots is increased, but whether this is the result of increased root growth or the cause of it is not clear. Losses of carbohydrates through respiration are reduced when the night temperature is lowered and it seems that respiration can be slowed down at night without impairing growth.

Alternation of day and night temperature is one of the factors affecting formation of tubers in potatoes. When they are growing in long days (16 hour light period) potato plants will not form tubers if the day temperature is 30 °C or more, whatever the night temperature, but at lower day temperatures tubers are formed if the night temperature is below day temperature. In short days (8 hour light period) maximum tuber formation occurs when both day and night temperatures are the same (optimum 17 °C). Tuber formation in potatoes is believed to be controlled by a specific tuber-inducing principle, the synthesis, transport or activity of which is affected by light and temperature.

4.2   Cell growth

Plants increase in size mainly as a result of cell enlargement. The rate of elongation of pea roots and maize coleoptiles increases almost linearly with increase of temperature between about 13 and 30 °C, and above this temperature the rate decreases sharply (Fig. 4–5).

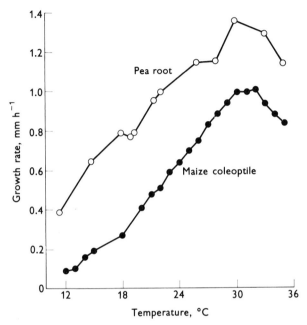

Fig. 4–5   Effect of temperature on the rate of elongation of pea roots (o——o) and maize coleoptiles (●——●). (Redrawn from LEOPOLD and KRIEDEMANN, 1975.)

Cell enlargement involves synthesis of cell wall constituents and cytoplasmic proteins as well as uptake of water and it is not surprising therefore that cell growth is sensitive to temperature. Water uptake is basically a physical process (see SUTCLIFFE, 1979) and so rapid osmotic changes have a low Q_{10}, but accumulation of solutes upon which prolonged water uptake depends is sensitive to temperature (see SUTCLIFFE and BAKER, 1974).

As cell enlargement is not unlimited, continued growth depends on production of new cells by division. Temperature affects both mitosis (the division of chromosomes) and cytokinesis (Fig. 4–6). In *Vicia faba* root meristems, both processes proceed very slowly at temperatures below 3 °C and most rapidly at about 25 °C.

Professor R. BROWN of the University of Edinburgh measured the rate of

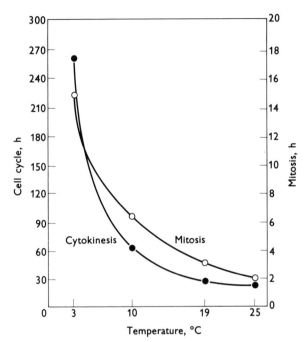

Fig. 4–6 Influence of temperature on the duration of mitosis (o——o) and the cell cycle (●——●) in *Vicia faba* root meristems. (Redrawn from STREET and ÖPIK, 1976.)

cell division and the duration of each phase of the mitotic cycle in pea roots at various temperatures and found that all stages were affected (Table 4). Differences between the effects of temperature on the various phases are no doubt attributable to differences in the temperature coefficient of the underlying chemical reactions. As temperature affects the rate of cell division in meristems it is not surprising that it also influences the rate of production of leaf primordia, but an effect of root temperature on leaf production in tobacco as shown in Fig. 4–7 when the shoots are kept at constant temperature, is unexpected.

In temperate regions, air temperature is the most important factor in renewed cambial activity in the spring. Initially the effect is attributable to an influence on bud-burst (see below), but during subsequent growth there is also a direct correlation between cambial cell division and temperature. An optimal spring temperature results in production of a larger number of xylem elements and a wider band of spring wood, providing that other factors such as water supply are not limiting.

Table 4   Effect of temperature on mitosis and cell division in *Pisum sativum* roots. Mean durations (in hours) are given for each stage of division at four temperatures (data from BROWN, 1953).

Stage	Mean duration (hours)			
	15 °C	20 °C	25 °C	30 °C
Interphase	22.60	17.00	14.50	13.30
Prophase	2.10	1.30	0.90	0.70
Metaphase	0.40	0.24	0.24	0.19
Anaphase	0.08	0.07	0.05	0.04
Telophase	0.37	0.22	0.19	0.16
Total	25.55	18.83	15.88	14.39
Q_{10}		1.60	1.30	

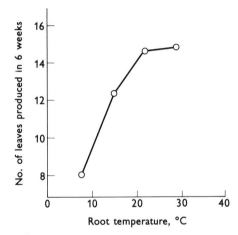

Fig. 4–7   Effect of root temperature on the number of leaves produced by tobacco plants (cultivar Hicks) in 6 weeks. (Redrawn from COOPER, 1975.)

4.3   Seed germination

Dry seeds are extremely tolerant of temperature and can withstand extreme cold or heat for a short time without ill-effects. Some have survived temperatures close to absolute zero (−273 °C) and at the other extreme, medick (*Medicago sativa*) seeds remain alive after heating to 120 °C for 30 minutes. However, high temperature does reduce the viability of seeds and it is best to store them at a low temperature.

There is a minimum and maximum temperature for every species of seed and outside this range germination will not occur. Barley, wheat and rye grains for example will not germinate below 3–5 °C or above 30–40 °C, while maize grains germinate only between about 8 °C and 44 °C. Sometimes the temperature at which germination will occur alters as the seed gets older. In *Brassica juncea*, only 8% of the seeds will germinate at 25 °C immediately after harvest but 3 weeks later 95% of the seeds germinate at this temperature.

Many seeds will grow perfectly well at a constant temperature but there are others, e.g. celery and gipsy-wort which germinate best at a fluctuating temperature (Fig. 4–8). The ecological significance of this response to a

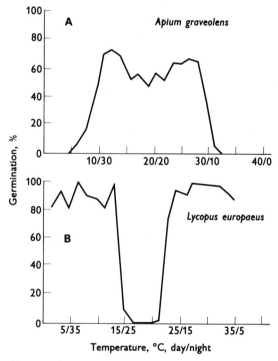

Fig. 4–8 Influence of diurnal fluctuation in temperature on germination of celery (*Apium graveolens*) and gipsy-wort (*Lycopus europaeus*). (From THOMPSON, 1974.)

diurnal alternation of temperature may be that it promotes the germination of those seeds that are close to the soil surface, where such fluctuations occur and suppresses that of more deeply-seated seeds. Why fluctuating temperature should stimulate seeds to germinate is still not clear but it has been suggested that alternation of temperature may increase the permeability of membranes and seed coats, to water and gases.

Germination of some seeds, e.g. lima beans and peas, is inhibited by brief periods of chilling to 5 °C or less during the imbibition phase. The effect is accompanied by a reduction in the rate of respiration which persists for some time. The effect of low temperature chilling on the germination of peas can be prevented by treatment of the seeds with a cytokinin during imbibition and this suggests that low temperature affects the balance between growth-promoting and inhibiting substances (see p. 34). In cotton seeds a brief exposure to low temperature during germination prevents conversion of stored lipid to carbohydrates which are required for growth. The effect does not seem to be exerted directly on the enzymes involved but rather on the membranes that separate the enzymes from their substrates. Possibly chilling causes a change in the structure of these membranes which affects their permeability (see p. 21).

Some seeds, notably those of rosaceous species (apple, cherry, peach, plum, etc.), will not germinate at all or only slowly until they have been exposed in a moist condition to low temperatures (0–10 °C) for several weeks or months (Fig. 4–9). The commercial practice of treating seeds in this way to prepare them for germination is called 'stratification' because the seeds are usually stored in layers. The requirement for exposure to

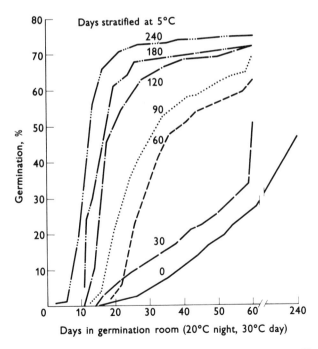

Fig. 4–9 Effect of period of stratification at 5 °C on germination of balsam fir (*Abies balsamea*) seeds. (Redrawn from DOWNS and HELLMERS, 1975.)

low temperature over an extended period before germination occurs prevents precocious growth in the autumn or winter in response to a brief warm period.

During stratification the embryo may grow, but in several cases no visible morphological changes occur in the seed. There are, however, profound changes in chemical composition. For example, in hogweed (*Heracleum* sp.) seeds, most of the dry matter is translocated from the endosperm into the cotyledons during stratification. There are also significant changes in some seeds in the distribution and proportions of various nitrogenous and phosphorus compounds. A reduction in the level of growth-inhibitory substances has been detected during stratification of apple seeds and in others, e.g. hazel seeds, an increase in growth-promoters such as cytokinins has been reported. In some (e.g. in peach) but not in all cases (e.g. apple) the cold-requirement for germination can be replaced by treatment with gibberellic acid which suggests that, sometimes at least, an alteration in the balance of growth promoters and inhibitors may act as a trigger for germination. (See VILLIERS (1975) for further information on this and the following subject.)

4·4 Bud dormancy

In temperate regions plants commonly become dormant before the onset of winter and growth occurs in the following season from resting buds. Dormant buds often develop in summer before the mean temperature begins to fall and the trigger is usually day-length. In many deciduous trees resting buds are formed in response to shortening days especially when these are accompanied by lower temperature or by water deficiency.

The breaking of bud dormancy may be controlled by photoperiod, temperature or, more usually, by a combination of the two. The buds of many trees and shrubs will not burst in the spring unless they have been exposed first to a prolonged period at low temperature (cf. stratification). Lilac twigs for example, if brought into a warm place in the autumn or early winter, will not begin to grow, but after several weeks' exposure to low temperatures the buds quickly burst when the temperature rises. Each bud behaves independently and if a lilac bush is kept in a glasshouse at a high temperature during the winter, with one branch protruding outside, the latter will begin to grow first when the temperature rises in the spring.

The requirement for a cold period to induce bud-break in fruit trees such as peaches sometimes limits the areas in which they can be grown. New varieties have been introduced which have a shorter chilling requirement and so can be cultivated successfully in places where the winters are less severe and the chance of spring frosts less likely.

There is an interesting effect of temperature on the breaking of dormancy in flower buds of some tropical species. It is well-known that flowering of coffee in Nigeria often occurs a fixed number of days after a shower of rain. Sometimes, e.g. after heavy rain, this can be attributed to a lowering of water stress but in other cases, where a light shower is effective, the effect is more probably due to the sudden lowering of temperature accompanying the rain. A similar explanation has been given for the regular flowering of *Dendrobium crumentatum* and other Malayan epiphytic orchids some 8–10 days after a thunder-shower following a dry spell.

Treatments that induce bud-break, sometimes, but not always, seem to cause a reduction in the levels of growth-inhibiting substances, and increases in the amounts of cytokinins and gibberellins in the buds have also been observed. In some cases, application of gibberellic acid will break bud dormancy, without a cold treatment, and it seems likely that, as in seeds, a change in the balance of growth-regulating substances is sometimes the trigger for bud-break.

4.5   Bulbs and corms

The life cycle of monocotyledonous plants such as daffodils, tulips and crocuses which flower in the spring or early summer and then die down producing dormant bulbs or corms which survive summer drought or winter cold, is strongly influenced by temperature.

The induction of bulbing in onions has been shown to be a photoperiodic phenomenon, but little is yet known about the control of bulb initiation in other species. Bulb and corm growth is of course strongly influenced by temperature and the induction of flowers is particularly affected. In some cases, e.g. narcissi, flower primordia are formed while the bulbs are still growing in size. However, in most bulbous irises the flower primordia are initiated during winter when the bulbs are dormant, but previous exposure to high temperature (20–30 °C) is required if flower formation is to occur at all. This is the reverse of 'vernalization' discussed in 4.6 below. Optimal temperatures for flower formation in narcissus and tulip bulbs is 17–20 °C, while for hyacinths 25 °C is best.

By manipulation of the temperature at which bulbs and corms are stored after lifting it is possible to regulate the time of flowering. Earlier than normal flowering of narcissi, hyacinths and tulips can be obtained by storing newly-lifted bulbs at fairly high temperatures, say 30–35 °C for a few days. Such 'prepared' bulbs can be made to flower several weeks earlier than normal in time for Christmas in northern latitudes. Such high temperature treatment has the added advantage of reducing the damage to flowers and foliage which may occur during subsequent treatment with hot water to control eel-worms. Immersion in water at 43–44 °C for 3–4 h is necessary to destroy eel-worm larvae in a medium-sized narcissus bulb and there is less risk of damage to the bulb by such treatment after a preliminary high temperature treatment (for other similar examples of thermal adaptation, see Chapter 6).

During the early growth of bulbs when new roots are being initiated and the shoot begins to elongate a low temperature (< 10 °C) is required. Anyone who has attempted to raise bulbs indoors in a warm place such as an airing cupboard knows the disastrous consequences! After formation of an adequate root system and emergence of the shoot, bulbs can safely be 'forced' to flower early by raising the temperature but it is best to do this gradually.

Sometimes, an unusually high temperature treatment is required to break bulb dormancy. In the South African Cape Province there are several rare bulbous species which remain dormant in the soil for years and begin to grow only after fire has swept over the area destroying the surrounding vegetation. In such cases it appears that a brief exposure to temperatures of several hundred degrees centigrade is required to initiate growth.

4.6   Vernalization

Some plants need to be exposed to a period of low temperature before they will flower. These include biennial plants, such as sugar beet, which normally remain vegetative in the first growing season and flower in the following year after prolonged exposure to cold. There are other plants, notably winter varieties of wheat and rye, which will flower if grown continuously at a high temperature, but do so more quickly after cold treatment. Thus by artificially exposing such plants briefly to a low temperature it is possible to make them flower earlier. This technique was applied by the Russian agriculturalist LYSENKO in the 1920s to the cultivation of cereals, and he called it 'jarovization', i.e. 'springization', which has been translated into 'vernalization'. The term is now applied not only to the artificial treatment of plants at a low temperature to promote flowering, but also to the process as it occurs in nature.

The seeds of some varieties of cereals known as winter annuals, e.g. winter wheat and Petkus rye, are normally sown in the autumn and begin to germinate at once. They survive the winter as young seedlings and during this time they are exposed to low temperatures. In the spring they continue their growth and flower in the early summer. Spring varieties of these cereals do not require low temperature treatment and can be sown in the spring. Winter varieties can also be planted at this time if they are treated at low temperature for a few days soon after germination. For most species temperatures in the range of 0–5 °C are optimal for vernalization but temperatures below zero are also effective if the seedling can withstand them. After vernalization immediate exposure to a high temperature nullifies the effect. In Petkus rye exposure of vernalized seedlings to 35 °C for even 24 hours immediately after the cold treatment is sufficient to cause devernalization. Petkus rye can be vernalized and devernalized several times in this way without ill-effects.

There is also an interaction between vernalization and photoperiod, at least in rye. Exposure of unvernalized grains to short days will substitute to a certain extent for the cold treatment. After vernalization, however, the flowering of Petkus rye is strongly promoted by long days.

Dormant seed cannot be vernalized as the low temperature stimulus is apparently received only by dividing cells in the young embryo or seedling. Once a meristem has been vernalized all buds derived from that meristem will produce early flowers. However, if there are some dormant meristems on a plant at the time it is vernalized, these will continue to produce flowers later as in an unvernalized plant. This seems to imply that vernalization, unlike the photoperiodic induction of flowering, does not produce a mobile 'florigen' or flower-producing substance, but rather causes a genetic modification, which alters the developmental pattern of individual groups of cells in the meristem.

Biennial plants typically grow vegetatively in the first season and form a rosette of leaves which may die down with the onset of winter leaving the plant to survive by a perennating organ such as a swollen root or hypocotyl. In the next spring a flowering shoot develops from the base of the plant and usually elongates rapidly, a phenomenon known as 'bolting'. Although bolting sometimes occurs in the first season it normally happens only after exposure of the plants for several days or weeks to temperatures close to o °C. Flowering in many biennials after the cold treatment is, as in winter annuals, promoted by long days and in some cases, e.g. the henbane (*Hyocyamus niger*) which has been much used in investigations of vernalization, flowering will not occur in short days. It has been observed that the low temperature requirement of some long-day plants, including henbane, can be satisfied by treating the apical meristem with gibberellic acid. However, gibberellins cause the flowering stem to elongate first and the flower buds are formed later whereas in a vernalized plant flower buds are formed before the stem elongates, so the role of gibberellins in the vernalization response is still uncertain.

A German plant physiologist, Melchers, claims to have demonstrated that the vernalization stimulus in biennial plants can be transmitted through a graft union. The basic finding is that a plant induced to flower by cold treatment will induce flowering in an unvernalized receptor plant on to which it is grafted. MELCHERS suggested that vernalization causes the production of a hormone 'vernalin' which induces flowering. However, such a substance has not yet been isolated or identified.

There are many species of plants which require cold treatment for flowering but which do not fall into either of the above categories. The chrysanthemum for example is a perennial plant flowering in short days, which requires a single cold treatment before it will respond to photoperiod. Cuttings taken from a plant which has been so treated remain responsive irrespective of the length of time since the cold treatment.

4.7 Thermotropism and thermonasty

If a temperature gradient exists across a growing organ the two sides will grow at different rates and a tropic curvature results. Usually the temperature gradient is too small to cause a significant effect but occasionally thermotropic responses have been reported in roots. When the organ bends towards the high temperature (growth stimulated on the cooler side) it is called +ve thermotropism, while bending towards the lower temperature is termed −ve (cf. phototropism and geotropism). Presumably the difference in behaviour is related to the optimal temperature for growth; if the higher temperature is above the optimum the organ will bend away from it and if it is below the optimum the response will be positive.

Nastic movements, such as alterations in the position of leaves, petals or flowers, occur in response to changes in an external factor, e.g. light intensity or temperature, without a gradient being involved. An example of thermonasty is the opening of tulip and crocus flowers when they are transferred from a lower temperature (say 5 °C) to a higher one (say 15 °C). If the flowers are cooled again the perianth returns to its original position and the opening and closing can be repeated as long as the petals continue to grow. A similar thing happens in the evening primrose (*Oenothera*) except that the flowers open when the temperature falls and close when it rises. These differences in behaviour are correlated with the behaviour of pollinating insects which is also regulated by temperature, so that the flowers are open when the insects visit them.

Flower movements are the result of differential growth on the two sides of the perianth segments. If the optimal temperature for growth of the upper surface is higher than for the lower, the upper surface may grow faster at this temperature than does the lower surface and this causes the flower to open. At a lower temperature the situation is reversed and the flower closes. On the other hand, if the upper surface grows faster at the lower temperature the flower will open when the temperature is lowered as happens in the evening primrose. It is said that tulip and crocus flowers will respond to changes in temperature of as little as 1 °C.

The flower stalks of many plants show thermonastic movements which result in the heads bending over at night to invert the flowers. This is particularly noticeable in some members of the *Umbelliferae*, e.g. in carrots and is said to be a device for preventing excessive loss of heat on cold nights. In such cases the two sides of the inflorescence stalk must respond differentially to temperature. The stalks fail to respond to change in temperature once the fruit has been set. The leaves of some plants also respond to falling temperature by undergoing so-called 'sleep movements' which cause a reduction in the exposed surface area and cut down the loss of heat at night. Such leaf movements occur, not by differential growth, but by reversible changes in turgidity of specialized 'motor' cells situated near the base of the leaf petiole or leaflet stalk. How variation in temperature can lead to differences in turgor between cells on the two sides of the stalk is not clear but it is likely that changes are induced in the permeability of the cell membranes to solutes causing an alteration in osmotic potential (see SUTCLIFFE, 1979).

4.8 Fruit ripening and storage

As might be expected, fruits tend to grow faster and ripen more quickly as temperature increases up to an optimum level, after which the growth rate gradually falls off. Dry fruits, e.g. cereal grains, and some 'succulent' fruits, such as grapes and oranges, ripen best when temperatures are high and fail to mature properly if it is too cold. The complex biochemical processes that are responsible for giving the characteristic flavour to fruits are very sensitive to temperature and the temperature that produces the best yield of fruit frequently does not give the best taste. The distinctive differences in flavour between apples grown in north temperate latitudes and those from further south is attributed largely to the lower temperatures occurring in the later stages of fruit ripening. Strawberry fruits, too, develop maximum flavour when grown at a lowish temperature (about 10 °C), while at 20 °C some varieties have hardly any taste at all, although they may grow larger and redden normally at the higher temperature. On the other hand, there are some fruits, especially of tropical species, which are injured by chilling even to 12 °C. Ripening bananas and melons are damaged by even a few hours' exposure to this temperature, which also adversely affects the growth of tomato and cucumber fruits (see Fig. 3–4).

Temperature has a marked effect on the rate of respiration of succulent fruits, especially those, e.g. apples, where gaseous exchange rises to a climacteric and then falls dramatically during ripening. Within certain limits the higher the temperature the sharper the rise and the higher the peak in oxygen consumption and carbon dioxide evolution. Low temperatures, on the other hand, tend to reduce or entirely suppress the climacteric. However, such treatment sometimes stimulates respiration as in citrus fruits, and this is thought to be an indirect effect induced by increased production of ethene (ethylene) at low temperature.

Because of the complexity of the effects of temperature on fruit ripening, the optimal temperature for storage varies markedly for different plants. Cereal grains and some other dry fruits can be stored safely at temperatures down to freezing point or lower but most succulent fruits are damaged by such treatment. Some varieties of apples, e.g.

Bramley's Seedless, can be stored for long periods at 3–5 °C without ill-effects but other apples brown prematurely at this temperature. It is not usually wise to store unripe fruits in a refrigerator with the intention of allowing them to ripen later as they often never ripen at all. It is better to keep them first at a higher temperature and transfer them to the refrigerator as soon as they are fully ripe.

Some fruits, e.g. pineapples, black and red currants and raspberries can be stored successfully for long periods in a deep freeze if they are frozen quickly when they are in the peak of condition. Unripe fruit will not ripen, either during or after such treatment, while over-ripe fruits become mushy as soon as they are thawed out.

5 Low Temperature Tolerance

5.1 Frost resistance

Susceptibility to injury by frost is probably the single most important factor limiting plant distribution, and it is also a major cause of damage to crops. Plants vary greatly in their sensitivity to low temperature; some tropical and sub-tropical plants are killed by chilling to temperatures still well above zero (see Chapter 3), whereas others, notably arctic-alpine plants, can survive prolonged exposure to sub-zero temperatures. Plants which can live at low temperatures and have a low optimal temperature for growth are termed *psychrophiles* ('lovers of frost'), although it is likely that they tolerate rather than enjoy such conditions.

Resistance to freezing injury alters markedly during the life of a plant. Most dry seeds, even of frost-sensitive plants, can survive temperatures that would kill a young seedling. Some dormant seeds have germinated after exposure for short periods to temperatures close to absolute zero (-273 °C). Resistance to freezing falls rapidly when the seed is hydrated and most germinating seeds are killed by temperatures of around 0 °C. During early growth many seedlings acquire a greater degree of low temperature tolerance if they are exposed gradually to low, but supra-lethal, temperatures. This is made use of in the horticultural practice of 'hardening-off' seedlings which have been raised at high temperatures in a glasshouse or frame. Once pea (*Pisum sativum*) seedlings have been kept at 5 °C for a few hours on several successive days they can withstand -3 °C without injury, whereas untreated plants would be killed. Hardening only occurs when plants are kept in the light and so products of photosynthesis are apparently necessary. However, there are some seedlings, e.g. those of the tomato, which cannot be hardened even in the light.

The freezing resistance of most plants changes with the season in step with changes in environmental temperature. In summer when temperatures are normally high many plants are more susceptible to a fall in temperature than they are in winter when it is colder. This process, which is known as *acclimatization*, is well illustrated by alpine plants. During the growing season, plants growing on Mt. Kurodake (1984 m, lat. 43.6° N, long. 143° E) in Hokkaido Province, Japan, are killed by freezing to -5°C to -7 °C and exposure to 0 °C to -3 °C for two weeks at this time increases the resistance to freezing only slightly. During the winter, however, most of the plants survive freezing to -30 °C and some, e.g. *Salix pauciflora*, mosses and lichens will withstand a temperature of -70 °C. The beginning of the development of freezing resistance in the late summer in

Salix species appears to be correlated with the development of dormant buds which occurs at different times in the various species during a period when air temperature is falling. A lowering of frost resistance happens in the spring at a time when air temperature is rising before the unfolding of new leaves (Fig. 5–1).

The dormant buds of woody plants are more frost resistant than the young unfolding leaves. It has been suggested that the protective scale-

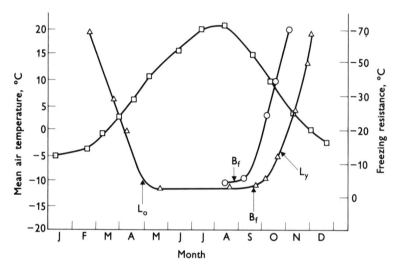

Fig. 5–1   Seasonal changes of freezing resistance in *Salix pauciflora* (o———o) and *Salix sachalinensis* (Δ———Δ) on Mt. Kurodake and mean air temperature (□——— □) throughout the year. L_o=leaves opening; L_y=leaves yellowing; B_f=buds forming. (Redrawn from SAKAI and OTSUKA, 1970.)

leaves of a bud help to retain heat generated by respiring cells (cf. Chapter 1) and so buffer any fall in temperature. However, the young foliage leaves are not metabolizing very actively at this time and it has not been possible with thermocouples to detect any significant rise in temperature as a result of such activity. In spite of the presence of surrounding scales the interior of a dormant horse chestnut bud (*Aesculus hippocastanum*) cools nearly as rapidly as the surrounding air when the temperature is allowed to fall from 20 °C to −20 °C and warms up again at a similar rate. However, scale leaves may help to reduce the cooling that occurs by radiation losses at night and as such they are particularly effective when they are dark in colour as for example in the ash (*Fraxinus excelsior*).

The below-ground parts of perennial plants, such as roots, tubers and rhizomes are less susceptible to frost damage than above-ground structures, partly because of a higher inherent frost resistance and also because radiation losses from underground organs are minimal, while at the same time soil temperature falls less quickly than that of the air.

5.2   Avoidance of freezing injury

Because plants do not have an effective built-in central heating system (see Chapter 1) they have to tolerate low temperatures rather than avoid them. Most plants, except those that are susceptible to chilling injury (Chapter 3), can tolerate freezing of water in the intercellular spaces without ill-effects, but neither frost-hardy nor tender plants can survive the formation of ice crystals within cells. The idea that the freezing of cell contents causes cells to expand and burst because of the increase in volume when water is converted to ice is a fallacy. Plant cell walls can easily stretch to accommodate the extra volume, especially as in the early stages of freezing when extracellular ice is formed, water is withdrawn from inside the cells with consequent loss of turgor. In fact it has commonly been observed that the volume of frozen cells is actually lower than that of normal turgid cells.

The key to frost-hardiness is the avoidance of intracellular freezing and some plants are remarkably effective in achieving this.

One of the ways in which the freezing point of plant sap may be depressed is by the accumulation of solutes (cf. the use of 'anti-freeze' in water-cooled automobile radiators). The temperature expressed in °C at which water freezes (T_F) at atmospheric pressure is related to osmotic potential (π) in pascals by the formula:

$$\pi = 12.22\, T_F$$

Thus a solution with an osmotic potential of -1222 kPa (-12.22 bar) has a freezing point of -1 °C. As the osmotic potential of plant cell sap rarely falls below -3000–4000 kPa (-30–40 bar) it is evident that accumulation of solutes is unlikely to cause a freezing-point depression of more than a few degrees and even in halophytes which often have very high concentrations of salt in their sap (but which incidentally are not noted particularly for their tolerance of freezing temperatures) depression of the freezing point to much below -10 °C is unlikely.

There are some indications that frost-tolerant wheat varieties tend to have higher sap concentrations than sensitive varieties, but no such correlation has been found in barley. In some plants, e.g. cabbage, it has been observed that the osmotic potential of the cell sap falls (i.e. it becomes more negative) during hardening of seedlings by exposure to low temperature, but storage of potato tubers at low temperature leads to conversion of starch to sugar without any reduction in susceptibility to frost.

In ivy (*Hedera helix*) leaves, water-soluble proteins in the cell sap increase gradually from late summer onwards but there is no sudden decrease in osmotic potential at the time when freezing tolerance increases sharply in the autumn, as might be expected if osmotic potential was important. Conversely, sap concentration remains high in the spring until after the time when sensitivity to low temperature has increased. Thus it seems likely that in several species at any rate sap concentration is of little significance in determining frost resistance.

An alternative explanation of frost tolerance is that a higher proportion of the water in hardy plants is bound to cell constituents and so does not freeze as readily as that in tender plants where it is mainly in a free state. In recent years the technique of nuclear magnetic resonance (NMR) has been used to study the state of water in plant tissues. This is a form of spectroscopy employing radiation at radio frequencies which makes use of the fact that there is a large spectral difference between liquid and bound or frozen water. However, attempts to correlate the quantity of unfreezable water with cold hardiness have been unsuccessful.

It is well known that although pure water melts at exactly 0 °C at atmospheric pressure, it does not freeze until it reaches a somewhat lower temperature, i.e. there is some degree of supercooling before freezing occurs. The water remains liquid until a microscopic ice crystal forms around a 'nucleus' after which the whole freezes rapidly. The temperature then rises as heat is liberated (latent heat of freezing) until it approaches 0 °C. Absolutely pure water in a clean vessel will freeze spontaneously without seeding but not until it reaches about − 40 °C. The presence of suspended material, e.g. dust particles, in the water causes nucleation to occur at a higher temperature, i.e. nearer 0 °C, than in pure water.

Some hardy plants seem to differ from tender ones in the degree of supercooling that occurs before ice crystals form. Various winter cereals, e.g. winter wheat, can withstand supercooling to − 25 °C without intracellular ice formation, and some woody plants do not form ice crystals in cells even at the temperature of liquid nitrogen (− 196 °C). The reason for this difference between plants is not yet quite clear but it appears to be related to the more effective isolation of extracellular ice crystals in frost tolerant plants. The plasma membrane and cell wall both act as barriers and it seems likely that the ability of the plasma membrane to maintain its structure at low temperature is an important fact in frost hardiness.

In a number of hardy woody species, e.g. *Buxus* and *Camellia*, ice crystal formation is localized in spaces between readily-separated layers of tissue in the leaf and is thereby rendered less likely to spread to other parts of the plant. Wilting of leaves at low temperature, as in *Rhododendron*, also appears to help to isolate extracellular ice crystals from intracellular contents.

There are a number of observations which indicate that the viscosity of protoplasm increases with frost hardening. Protoplasmic viscosity is 'structural' or non-Newtonian because protoplasm is not a simple liquid.

SCARTH and LEVITT (1937) noticed that many cells of hardy plants show convex plasmolysis whereas those of tender cells more often show concave plasmolysis (Fig. 5–2). They attributed this to a difference in the consistency of protoplasm arising from a change in the structure of the constituent proteins. A possible mechanism relating protein structure to frost resistance is discussed below (5.4).

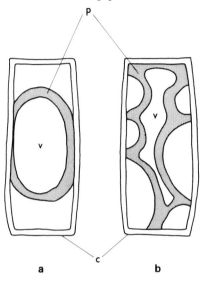

Fig. 5–2   Types of plasmolysis: (a) convex (b) concave. c=cell wall; p=protoplasm; v=vacuole. (From SUTCLIFFE, 1979.)

5.3   Effects of mechanical stress

On the basis of many observations and experiments the German plant physiologist, ILJIN (1933), proposed that freezing tolerance is related to mechanical stress. He observed that cells often collapse when extracellular ice formation occurs and concluded that the protoplasm of such cells is placed under mechanical stress which causes injury. Any factor which reduces such stress during freezing or thawing would be expected to increase hardiness. The smaller the cell size, the greater the specific surface area and the less the volume strain per unit surface for any degree of contraction. It has been observed in fact that frost hardiness is often correlated with small cell size. Also the greater the concentration of solutes, the less the water loss and therefore the less the cell contraction and mechanical stress at any one temperature. An increase in the amount of bound water would have the same effect. A great many observations indicate that low temperature tolerance is directly related to a low degree of cell contraction as a result of extracellular freezing.

Another factor that appears to be correlated with frost hardiness is the degree of dehydration of the protoplasm when water is withdrawn from the cell. If protoplasm is dehydrated it becomes stiffer and more likely to be damaged by mechanical stress. In hardened plants protoplasmic dehydration is minimal at a given temperature. The Russian, TUMANOV (1967) has suggested that when plants cells harden the cell contents change from sol to gel and this renders the protoplasm less liable to damage by mechanical deformation or dehydration. A slow decrease in the freezing point of the solution in the gel lattice as it increases in concentration protects the cells from ice formation. Dissolved substances are thought to act not only as an antifreeze but as a plasticizer of the gel. TUMANOV also concludes that there are changes in the structure of the plasma membrane which protects them from damage when dehydration occurs (see below).

5.4   Levitt's hypothesis

The American plant physiologist J. LEVITT has proposed that freezing injury is caused by aggregation of proteins as a result of the formation of —S—S— (disulphide) bonds from —SH (sulphydryl) groups in adjoining protein molecules as they approach one another when the protoplasm is freeze-dehydrated (Fig. 5–3). The protein molecules thus undergo a conformational change which results in destruction either of enzyme activity or membrane integrity. Low temperature tolerance can therefore be attributed to prevention of disulphide bonding.

The —SS— ⇌⊢—SH hypothesis is based on the observation that an

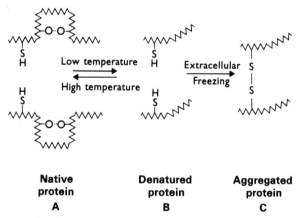

Native protein   **Denatured protein**   **Aggregated protein**

A   **B**   **C**

Fig. 5–3   Levitt's hypothesis. **A–B**, how temperature causes reversible denaturation of protein by weakening hydrophobic bonds (o – – – o). **C**, extracellular freezing causes dehydration bringing —SH groups closer together when aggregation occurs with formation of —S—S— bonds. (Redrawn from LEVITT, 1969.)

increase in the number of protein —SS— groups accompanies freezing injury but does not occur when freezing does not lead to injury. Furthermore there is a correlation between the —SH content of proteins in a tissue homogenate and frost tolerance. The proteins from frost tolerant plants have fewer —SH groups and they are less readily oxidized than those of tender plants.

This hypothesis explains a number of puzzling features of frost tolerance, one being the relationships between growth and sensitivity to low temperature. Rapid growth is associated with high —SH content and also with sensitivity to frost damage. Naturally, the larger the number of —SH groups per unit protein the more difficult it must be to prevent their oxidation to intermolecular —SS— bonds on freezing.

The hypothesis has been tested in a model system with a thiolated form of gelatine, thiogel. The melting point of a gel made from thiogel rises with an increase in the number of —SS— bonds so this can be used as an indication of —SS— bond formation. When the gel is freeze-hydrated its melting point rises much more rapidly than when it is kept unfrozen. Because thiogel is a denatured protein it cannot be used to test the hypothesis that denaturation follows intermolecular binding. Experiments with a native protein, bovine serum albumin (BSA), have confirmed that the native —SS— protein does not aggregate on freeze dehydration but the —SH protein does.

On the basis of experiments with model proteins, LEVITT suggests the frost injury results from the following chain of events:

1. Low temperature denatures proteins reversibly unmasking reactive —SH groups.
2. Freeze dehydration (a) removes vacuolar water producing cell contraction which applies stress on protoplasmic proteins activating their —SH bonds and (b) removes protoplasmic water decreasing the distance between the reversibly denatured protein molecules.
3. Intermolecular bonding due to —SH oxidation to form —S—S— bonds aggregates proteins irreversibly, killing the cell. The scheme may be summarized as follows:

$$\begin{array}{ccc} \text{low temperature} & \text{low temperature} & \\ \text{N} \rightleftharpoons & \text{D} \longrightarrow & \text{A} \\ \text{optimal temperature} & & \end{array}$$

where N=native protein; D=denatured protein; A=aggregated protein (see Fig. 5–3).

The instantaneous nature of freezing injury suggests that membrane damage is responsible rather than coagulation of soluble enzymes. It has been observed that plasma membrane and mitochondrial proteins in animals and microorganisms have high —SH contents. It is likely that membrane proteins are unfolded at low temperatures by weakening of hydrophobic bonds (which are temperature sensitive) making it possible for first, reversible denaturation, and then irreversible aggregation to occur. This may cause the formation of permeable non-lipid holes in the lipid layer of the plasma membrane. It is suggested that extracellular freezing causes cell contraction and therefore a tension on the surface (Fig. 5–4). When the tension becomes sufficiently severe a break occurs in

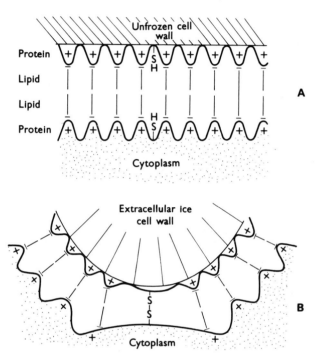

Fig. 5–4 Diagrammatic representation of a possible mechanism of membrane protein aggregation due to cell collapse on freeze dehydration. **A**, normal state; **B**, cell wall collapsed. (Redrawn from LEVITT, 1972.)

the thin bimolecular lipid layer and the two layers of membrane protein come into contact with each other and this contact results in covalent bonding as a result of which the hole becomes permanent leading to efflux of cell contents on thawing and death. Freezing injury thus resembles chilling injury in that the effect is primarily on membrane structure but whereas chilling causes a lipid phase change (see Chapter 3) LEVITT proposes that freezing has its major effect on the state of the protein component. Whether or not this hypothesis is correct remains to be established but it seems likely to me that membrane disruption is important in frost injury.

6 High Temperature Tolerance

6.1 Thermophiles

Most plants, when they are actively growing, cannot survive for long at temperatures in excess of about 40 °C. Those which can tolerate higher temperatures are termed *thermophiles* ('heat-lovers') and they include the inhabitants of hot deserts, tropical forests and hot springs. There have been reports that some blue-green algae can survive, and even grow, at temperatures as high as 93–98 °C while some thermophilic bacteria continue growing up to 100 °C and are probably killed only when the water they contain is turned to steam. The highest recorded temperature for growth of a higher plant is about 65 °C for a cactus (*Opuntia* sp.) but most thermophiles cannot tolerate prolonged exposure to temperatures in excess of about 55 °C. It has been found that the temperature at which 50% of the plants were killed in some 39 species on the coast of Spain ranged between 44 °C and 55 °C during the months of August and September.

In contrast to the relatively minor significance of exposure time in the case of freezing injury, the time for which plants are subjected to high temperature is of great importance in relation to survival. In general, heat-killing temperature varies inversely with exposure time and according to LEPESCHKIN (1912):

$$T = a - b \log Z$$

where T is the killing temperature; Z the time of exposure (in minutes); a and b are constants which differ for different species. The values obtained from this equation when $a = 79.8$ and $b = 12.8$ agree closely with the measured values for the temperature at which coagulation occurs in *Tradescantia discolor* leaf cells (Table 5).

The relationship between heat-killing temperature and time is exponential so that Arrhenius plots of logarithms of heat-killing time and the reciprocal of absolute temperature should give straight lines. A change in slope in the Arrhenius plot of survival in Canadian Pond Weed (*Elodea canadensis*) (Fig. 6–1) at about 50 °C indicates that there are two distinct phases in thermal injury (cf. chilling injury—Chapter 3).

6.2 Sensitivity to heat injury

As with freezing injury, dormant tissues in a dehydrated state are much less sensitive to high temperature injury than are actively-growing plants.

Table 5   Death temperature of *Tradescantia discolor* leaf cells for different heating times determined by direct observation of coagulation and by calculation (from LEPESCHKIN, 1912).

Heating time (min)	Death temperature (°C)	
	Determined	Calculated
10	69.6	67.0
25	63.2	62.0
60	57.0	57.1
80	55.7	55.5
100	54.1	54.2

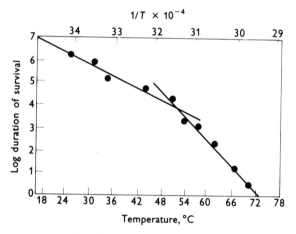

Fig. 6–1   Arrhenius plot of survival time of *Elodea canadensis* at various temperatures. (From BELEHRADEK and MELICHAR, 1930.)

Many moderate thermophiles and even mesophiles survive during the hottest periods of the year as seeds or dormant vegetative organs such as bulbs and corms. Some hard-coated seeds that do not take up water readily can be autoclaved at 120 °C for 30 minutes without being killed, but they do not survive for long if the seed coat is damaged. As seeds take up water their heat resistance falls dramatically. Dry wheat grains can withstand a temperature of 90 °C for up to about 10 minutes but after soaking for 24 hours, 60 °C kills them in about 1 minute.

Young seedlings are often very sensitive to high temperatures and the region of the stem or hypocotyl close to the soil surface is often the most susceptible. In mature plants heat injury in nature has most often been reported in bulky organs such as fleshy fruits, or in the trunks of thin-barked trees where the cambium may be damaged, by exposure to high insolation, but rarely in leaves, presumably because of the cooling effect of transpiration. It has been found with plants growing in the Sahara Desert that leaf temperature commonly rises to within a few degrees of the lethal temperature, and some species survive only if sufficient water is available to maintain high transpiration rates (see Chapter 2).

In some cases there is an increase in the heat tolerance of leaves during hot periods of the year. It is claimed that the cross-leaved heath (*Erica tetralix*) would not survive in warmer parts of Europe if it were not for a rise of 6–7 °C in heat tolerance of the leaf tissue during the summer.

6.3   Causes of heat injury

6.3.1   Injury by desiccation

High temperature may injure a plant indirectly by causing it to dry out. Transpiration rate increases with increase in temperature because of a direct effect on the diffusion coefficient of water and also because there is an increase in the water potential (vapour pressure) gradient between the plant and surrounding air (see SUTCLIFFE, 1979). If the relative humidity of the air is 70% it may be calculated that a 5 °C rise in leaf temperature above that of the atmosphere will double the transpiration rate. Closure of the stomata to conserve water may only aggravate the situation because of the decreased cooling effect of evaporation (see Chapter 2).

Most desert thermophiles are xerophytes, i.e. they are morphologically and physiologically adapted to conserve water at times of water stress. This is achieved by such adaptations as an impermeable cuticle, succulent habit and sunken stomata which open only at night. Xerophytism must be coupled with high heat tolerance because with lowered transpiration rates the temperature of the illuminated plant will inevitably rise; death by heat is no less final than death by desiccation.

In tropical rain forests the problem is somewhat different because transpiration is often limited by high humidity, and drought is not a serious danger. Light intensities within a forest canopy are also much lower than in the open and therefore the heating effect of insolation is reduced. Temperatures are much more uniform throughout the year and the plants of rain forests rarely if ever suffer from heat injury.

6.3.2   Effects on metabolism

High temperature may cause injury to a plant because of its differential effect on various metabolic processes. As has already been mentioned (p. 26) respiration usually has a higher temperature optimum than photosynthesis, so as temperature rises the balance between the two processes is disturbed. The temperature at which the two proceed at the same rate is called the *temperature compensation point* (TCP). Above this temperature respiration is more rapid than photosynthesis and the food reserves eventually become exhausted, leading to starvation and death of the plant.

Thermophiles have such a high TCP that starvation is probably not a common cause of injury even at temperatures near to their limit, but it may account for decrease in growth rate and eventual death of some mesophiles growing at temperatures just above their TCP, especially at low light intensity. This may also explain why some plants cannot withstand such high temperatures at night as they can during the day (see p. 27).

Rapid respiration induced by high temperature, especially at night, may lead to a shortage of oxygen in the tissue and the accumulation of toxic products of anaerobic respiration. This could account for the fact that the temperature at which injury occurs can sometimes be increased, or the time of exposure necessary for damage lengthened, by raising the oxygen concentration. In seedlings of *Pennisetum typhoides*, toxic ammonia levels occur after exposure to 48 °C for 12–24 hours indicating that high temperature can also cause disruption of nitrogen metabolism.

There is evidence that the failure of plants to grow properly above a critical temperature is sometimes due to failure in the synthesis of an essential metabolite. Such a biochemical lesion can be off-set by supplying the missing substance. Thus a mutant of the fungus *Neurospora crassa*, which can be grown on a standard medium only up to about 25 °C, was induced to grow at 35–40 °C by supplying riboflavin in the medium. Similarly, the growth of two races of the crucifer, *Arabidopsis thaliana*, at a supra-optimal temperature of 31.5 °C was increased when the medium was supplemented with biotin, while a third race responded to cytidine (LANGRIDGE and GRIFFING, 1959).

The effect of high temperature in producing a biochemical lesion may be either to increase the rate of breakdown or inhibit synthesis of an essential constituent of the organism, whether it be a vitamin, enzyme or structural component. Heat injury will occur at a temperature where the rate of resynthesis of the substance is insufficient to compensate for its degradation. There is some evidence that thermophilic bacteria can synthesize various enzymes more rapidly at a given temperature than can mesophiles. In addition, a variety of enzymes have been found to be more thermostable when extracted from a thermophile than from a mesophile. AMELUNXEN and LINS (1968) compared the heat tolerance of eleven enzymes extracted from *Bacillus stearo-thermophilus* (a thermophile) with the same enzymes from *Bacillus cereus* (a mesophile) and found that with two exceptions the enzymes from the former were more stable. In a higher plant, the reed-mace (*Typha latifolia*), MCNAUGHTON (1966) found that malate dehydrogenase extracted from leaves of an ecotype native to a hot climate was much more resistant to heating than the same enzyme extracted from ecotypes adapted to cooler climates. On the other hand two other enzymes, a transaminase and an aldolase, showed the same heat tolerance regardless of origin. It may be concluded from such observations that some, but not all, enzymes in thermophiles are more thermostable than their counterparts in mesophiles. Some, e.g. aldolase from reed-mace, seem to be sensitive to temperature even when extracted from thermophiles and this suggests that the enzyme may be more heat-stable inside the cell than after extraction.

Injury may result from a net breakdown of protoplasmic proteins at high temperatures. It has been found that the yellowing of some leaves (e.g. of tobacco) that occurs after brief exposure to a temperature of about 50 °C can be prevented by treatment with kinetin. This has been attributed to a stimulation of protein synthesis which offsets increased proteolysis at the high temperature.

6.3.3 Protein denaturation

The effects of high temperatures on metabolism discussed in 6.3.2 are likely to produce visible injury or death only after a period of time. Sometimes, however, high temperature has an immediate effect, e.g. on growth, or on protoplasmic streaming. Such effects are often attributed to a direct effect of temperature or structural proteins in the cytoplasm. ALEXANDROV (1964) found that cytoplasmic streaming is halted at a lower temperature than other cell processes and concluded that the proteins involved in protoplasmic streaming are more sensitive to heat denaturation than are other cytoplasmic proteins.

According to LEVITT (1969) heat injury is due to protein denaturation at high temperature followed by aggregation thus:

$$N \underset{\text{optimal temperature}}{\overset{\text{high temperature}}{\rightleftharpoons}} D \overset{\text{high temperature}}{\longrightarrow} A$$

where
N = native
D = denatured
A = aggregated protein molecules
(cf. low temperature injury, p. 47).

As hydrophobic bond strength increases with temperature, denaturation must be brought about mainly by breakage of hydrophilic bonds the strength of which is not so affected by temperature. Thermophiles have been found to have a higher proportion of hydrophobic to hydrophilic groups in their proteins and this would account for their greater stability at high temperature. Direct heat injury is likely to be due to denaturation and aggregation of membrane proteins causing an increase in cell permeability and death (cf. Fig. 5–4).

6.4   Heat-hardening

The fact that the heat resistance of plants changes with the season (see p. 41) suggests that mesophiles can be hardened by exposure to a high temperature, just as psychrophiles can be adapted to withstand low temperatures. However, early experiments designed to test this gave the opposite results. Exposure to high sub-lethal temperatures for extended periods actually reduced heat tolerance while low-temperature hardening increased it. It is now clear that this effect was caused by a depletion of carbohydrate reserves by the high-temperature pretreatment and an accumulation of reserves at the low temperature, which may have masked a direct hardening effect. More recently, there have been a number of reports that very brief exposure to moderately high temperature can increase the heat resistance of a plant. In addition it has been shown that various enzymes extracted from such heat-hardened plants are significantly more heat stable than when extracted from unhardened plants. LEVITT believes that heat-hardening causes a conformational change in the proteins which results in a higher proportion of hydrophobic bonds and increased thermostability.

It has been found that heat-hardening does not cause frost tolerance, but rather the reverse. This is to be expected since more highly hydrophobic proteins should be less stable at low temperature (see p. 46). On the other hand, when a plant becomes frost-hardened in the autumn its heat tolerance also increases. This correlation is puzzling since heat tolerance cannot be of any significance physiologically during the winter. The explanation may be that the changes induced by low temperature which prevent aggregation of denatured protein also prevent aggregation at high temperature.

Further Reading

General

MERYMAN, H. T. (ed.) (1966). *Cryobiology*. Academic Press, New York and London.
ROSE, A. H. (ed.) (1967). *Thermobiology*. Academic Press, New York and London.

Chapter 1

HALL, J. A. (1953). *Fundamentals of Thermometry*. Institute of Physics, London.
HALL, J. A. (1953). *Practical Thermometry*. Institute of Physics, London.
JANZEN, D. H. (1975). *Ecology of Plants in the Tropics*. Studies in Biology no. 58. Edward Arnold, London.
PRUITT, W. O. (1978). *Boreal Ecology*. Studies in Biology no. 91. Edward Arnold, London.
TRANQUILLINI, W. (1964). The physiology of plants at high altitudes. *A. Rev. Pl. Physiol.*, **15**, 345–60.

Chapter 2

GATES, D. M. (1962). *Energy Exchange in the Biosphere*. Harper and Row, New York.
GATES, D. M. (1965). Heat transfer in plants. *Scient. Am.*, **213**, 76–84.
RASCHKE, K. (1960). Heat transfer between the plant and the environment. *A. Rev. Pl. Physiol.*, **11**, 111–26.

Chapter 3

RAISON, J. K. (1973). The influence of temperature-induced phase changes on the kinetics of respiratory and other membranes-associated enzyme systems. *Bioenergetics*, **4**, 285–309.

Chapter 4

VEGIS, A. (1963). Climatic control of germination, bud-break and dormancy. In: *Environmental Control of Plant Growth*. Ed. L. T. Evans. Academic Press, New York and London, 265–87.
WENT, F. W. (1970). In: *Plant Agriculture*, Readings from *Scientific American*. Freeman, San Francisco, 108–18.

Chapter 5

BURKE, M. J., GUSTA, L. N., QUANN, H. A., WEISER, C. J. and LI, P. H. (1976). Freezing and injury in plants. *A. Rev. Pl. Physiol.*, **27**, 507–28.
OLIEN, C. R. (1967). Freezing stresses and survival. *A. Rev. Pl. Physiol.*, **18**, 387–408.
WARREN-WILSON, J. (1966). An analysis of plant growth and its control in arctic conditions. *Ann. Bot.*, **30**, 383–402.

Chapter 6

BROCK, T. D. (1967). Life at high temperatures. *Science*, **158**, 1012–19.
LANGRIDGE, J. (1963). Biochemical aspects of temperature response. *A. Rev. Pl. Physiol.*, **14**, 441–62.

References

ALEXANDROV, V. Y. (1964). Cytophysical and cytoecological investigations of heat resistance of plant cells towards the action of high and low temperature. *Q. Rev. Biol.*, **39**, 35–77.

AMELUNXEN, R. and LINS, M. (1968). Comparative thermostability of enzymes from *Bacillus stearothermophilus* and *Bacillus cereus*. *Archs biochem. Biophys.*, **125**, 765–9.

BELEHRADEK, J. and MELICHAR, J. (1930). L'action différente des températures élevées et des températures normales sur la survie de la cellule végétale (*Helodea canadensis Rich.*). *Biologia Gen.*, **6**, 109–24.

BLACKMAN, F. F. (1905). Optima and limiting factors. *Ann. Bot. (Old Series)* **19**, 281–95.

BROWN, R. (1953). The effects of temperature on the durations of the different stages of cell division in the root tip. *J. exp. Bot.*, **2**, 96–110.

COOPER, A. J. (1973). *Root Temperature and Plant Growth*. Research Review, **4**, Commonwealth Bureau of Horticultural and Plantation Crops, East Malling, Kent.

DOWNS, R. J. and HELLMERS, H. (1975). *Environment and Experimental Control of Plant Growth*. Academic Press, London and New York.

GATES, D. M. (1968). Transpiration and leaf temperature. *A. Rev. Pl. Physiol.*, **19**, 211–38.

HAAS, P. and HILL, T. G. (1929). *An Introduction to the Chemistry of Plant Products*. Vol. II, *Metabolic Processes*. Longmans, Green, London.

ILJIN, W. S. (1933). Über den Kältetod der Pflanzen und seine Ursachen. *Protoplasma*, **20**, 105–24.

LANGRIDGE, J. and GRIFFING, B. (1959). A study of high temperature lesions in *Arabidopsis thaliana*. *Aust. J. biol. Sci.*, **12**, 117–35.

LEMON, E. (1963). Energy and water balance of plant communities. In: *Environmental Control of Plant Growth*. Ed. L. T. Evans. Academic Press, New York and London, 55–78.

LEOPOLD, A. C. and KRIEDEMANN, P. E. (1975). *Plant Growth and Development*, 2nd Edition. McGraw-Hill, New York.

LEPESCHKIN, W. W. (1912). Zur Kenntnis der Einwirkung supramaximaler Temperaturen auf die Pflanzen. *Ber. dt. bot. Ges.*, **30**, 703–14.

LEVITT, J. (1969). Growth and survival of plants at extremes of temperature—a unified concept. *Soc. exp. Biol. Symp.*, **23**, 395–448.

LEVITT, J. (1972). *Responses of Plants to Environmental Stresses*. Academic Press, New York and London.

LYONS, J. M. (1973). Chilling injury in plants. *A. Rev. Pl. Physiol.*, **24**, 445–66.

LYONS, J. M. and RAISON, J. K. (1970). Oxidative activity of mitochondria isolated from plant tissues sensitive and resistant to chilling injury. *Pl. Physiol.*, **45**, 386–9.

MCNAUGHTON, S. J. (1966). Thermal inactivation properties of enzymes from *Typha latifolia* L. ecotypes. *Pl. Physiol.*, **4**, 1736–8.

MINCHIN, A. and SIMON, E. W. (1973). Chilling injury in cucumber leaves in relation to temperature. *J. exp. Bot.*, **24**, 1231–5.

PEARSALL, W. H. (1950). *Mountains and Moorlands*. Collins, London.

SAKAI, A. and OTSUKA, K. (1970). Freezing resistance of alpine plants. *Ecology*, **51**, 665–71.

SCARTH, G. W. and LEVITT, J. (1937). The frost-hardening mechanism of plant cells. *Pl. Physiol.*, **12**, 51–78.

STREET, H. E. and ÖPIK, H. (1976). *The Physiology of Flowering Plants*, 2nd Edition. Edward Arnold, London.

SUTCLIFFE, J. F. (1979). *Plants and Water*. Studies in Biology no. 14, 2nd Edition. Edward Arnold, London.

SUTCLIFFE, J. F. and BAKER, D. A. (1974). *Plants and Mineral Salts*. Studies in Biology no. 48. Edward Arnold, London.

SWEENEY, B. M. (1969). *Rhythmic Phenomena in Plants*. Academic Press, London and New York.

TANNER, C. B. and LEMON, E. R. (1962). Radiant energy utilized in evapotranspiration. *Agron. J.*, **54**, 207–12.

THOMPSON, P. A. (1974). Effects of fluctuating temperatures on germination. *J. exp. Bot.*, **25**, 164–75.

TUMANOV, I. I. (1967). Physiological mechanisms of frost resistance of plants. *Fiziologiya Rast.*, **14**, 520–39.

VILLIERS, T. A. (1975). *Dormancy and the Survival of Plants*. Studies in Biology no. 57. Edward Arnold, London.

WENT, F. W. (1945). Plant growth under controlled conditions. V. Relation between age, variety and thermoperiodicity of tomatoes. *Am. J. Bot.*, **32**, 469–79.

WENT, F. W. (1957). *The Experimental Control of Plant Growth*. Chronica Botanica Co., Waltham, Mass.